Mrs Harry Hu
The Lady Globetrotter

The Story of a Woman's Endurance

By Robert Hamilton

ISBN No 9798709238145

Copyright © Robert Hamilton 2021

All rights reserved. No part of this book may be reproduced or transmitted in any form or by any means, electronic or mechanical including photocopying, recording or by any information storage and retrieval system without permission from author.

Google and Google Maps are trademarks of Google LLC and this book is not endorsed by or affiliated with Google in any way.

Cover designed by David Millichope

The Lady Globetrotter

The Lady Globetrotter

Elizabeth Yates, The Lady Globetrotter, taken in Luddenden, Yorkshire

The Lady Globetrotter

Foreword

I like to think of myself as a bit of an amateur historian, with a particular interest in the First World War, or the Great War as it became known. In 2012 I joined the Halifax Great War Heritage Society and, as part of one of the society projects, I was browsing through a copy of our local newspaper, the Halifax Weekly Courier, for August 1914, to see how the outbreak of war was recorded. As I am sure anybody who has looked through old newspapers for particular items of information will tell you it is very easy to get side tracked, and that is what happened to me. I read an article about Mrs Harry Humphries, The Lady Globetrotter, a local woman who was walking around the world and I was immediately intrigued. Who was she, why was she attempting such an extraordinary venture and what happened to her?

Further research followed, and, from the excellent archive of historic newspapers which have been digitised by the American Library of Congress, I found a number of articles relating to some of the adventures of Mr and Mrs Harry Humphries as they trudged their way around the world. Census records from this country allowed me to trace the family of Mrs Humphries and her husband Harry. But there were still some significant gaps in the story, so I placed an appeal for information in the local newspaper and on the website of the Halifax Great War Heritage Society and got absolutely no response. I had hit a dead end.

Two years later, completely out of the blue, I received an email from the great nephew of Mrs Humphries who told me he had hundreds of items about the Lady Globetrotter and I was welcome to see it if I wished. Naturally I did and found an absolute wealth of material including items of clothing worn by the lady herself, photographs, scrapbooks, various items of documentation, an actual cine film of the Globetrotters entry into Dortmund, Germany and, best of all, a series of letters written by the Lady Globetrotter to her family which gave valuable insights into her life and events on her journey. All this led to additional research and the more I found out the more questions were posed, some of which remain unanswered to this day.

If you have any further information about the 'Lady Globetrotter' or her life please contact me at :-

robhamilton1914@gmail.com

Many people have contributed directly and indirectly into the writing of this book. In particular I would like to thank David Millichope for his unfailing support, my wife Linda who tolerated my absences from family life and above all David Marcer, the great nephew of the Lady Globetrotter, and his wife Susan for their hospitality and for allowing me complete access to the Globetrotter archive. But for their presence of mind the archive may well have been lost.

Robert Hamilton
Halifax, West Yorkshire
2021

Chapter 1
A Beginning

Like all good stories it is always best to start at the beginning. Mrs Harry Humphries, later to be known as 'The Lady Globetrotter', was born Elizabeth Ann Yates at South View Terrace, Kelsey Street, Halifax, Yorkshire on 21st April 1883, the eldest child of engineer Alfred and farmer's daughter Hannah (nee Kelsey). Alfred's family had moved to Halifax sometime in the 1860s, no doubt attracted to the town by its growing engineering industry. Halifax born Hannah's parents lived in the nearby unusually named Esps Farm and perhaps they had built the houses on Kelsey Street on their land, hence its name. The reason has been lost to time. Five more children followed, Ethel Kelsey 1885 (died in infancy), Alfred Theodore 1886, Beatrice Olive 1890 (died in infancy), Oswald 1892 and Frederick Kelsey 1899. Sometime between the birth of Alfred, who was known as Theo to the family, and the census of 1891 the family had moved to the small village of Luddenden, a few miles outside Halifax, where Alfred set up an engineering business in the Old Corn Mill in the village.

Elizabeth was born into a time when the Victorian era and the Empire were at their height. The country had a stable government, although much of the male and all of the female population were unable to vote. Victorian society was organised hierarchically, mainly based around gender and class. Great Britain was the wealthiest country in the world, although that wealth was not fairly shared, and there was a burgeoning middle class which would include Lizzie's family due to their father's status as an engineer and business owner, and the income that entailed. In this hierarchy women were dependent on the men in their lives, be they father, husband or brother. Men were meant to participate in politics and be the family's breadwinner while women were meant to run households and raise families. For an independent and adventurous spirit such as Elizabeth this must have been stifling. Her family recollect that she always wished to go on the stage but was forbidden to do so by her father and in a later letter she mentions how she wrote stories which he would tear up if he found them.

We do not know much about Elizabeth's childhood. She was known as Lizzie to the rest of the family although she was later also known to use the name Elsie. She and her brother Theo were regarded as sickly children and spent time in a sanatorium, believed to be The Ormerod Convalescent Home at St Anne's on Sea, Fylde, Lancashire. Both joined the Health & Strength League, an organisation whose objective was, "to unite together all Physical Culturists throughout the world for the purpose of disseminating the broad principles of, and in Nature's way promoting, the cause of Health and Strength". Theo took up weightlifting and Lizzie the more genteel and feminine pastime of walking, she claimed to have climbed Mount Snowden, the highest mountain in Wales, on several occasions. This was no mean feat, Sir Edmund Hillary, the first man to conquer Everest, climbed the mountain as part of his training for the attempt. Even to this day there are strict warnings about the changeable weather and the need to wear suitable clothing and equipment.

In October 1904 tragedy struck the family when, after a long struggle, Lizzie's mother, Hannah, died of cancer at the age of 54. The funeral service took place in St James' Free Church, Luddenden where the family had been actively involved in raising funds for the building of the new church. Hannah had been given the honour of laying one of the corner stones. She was interred in the family grave at Christ Church, Pellon alongside her parents and other family members. In her wonderfully detailed will she left Lizzie her gold watch and chain, her diamond ring, the family bible, her sewing machine and a legacy of £25. According to the social customs of the time, Lizzie would have been expected to act as housekeeper and surrogate mother to her three younger brothers although Theo reminisced many years later that,

"Lizzie was always something of a wanderer, never much good at housework".

Then in June 1908 there was more upheaval. Lizzie emigrated to Winnipeg, Canada - alone, a brave and daring thing to do for a woman in the early 20th Century. It is not known what prompted this drastic move. There is a possible clue in the collection of family memorabilia which contains a faded and creased photograph of Lizzie and a rather dapper young man, she in her Sunday best and he in top hat and tails. On the rear is written *'Lizzie and her fiancé taken in Wesley House, Luddenden'*, no other details are given and the identity of the fiancé is unknown. Was Lizzie's abrupt departure prompted by the breakdown of the relationship with her fiancé or was she perhaps escaping from the restricted role of women in the suffocating Victorian class and social structure of the time. We shall probably never know.

Whatever the motivation Lizzie went to Canada to seek fame and fortune, but little is known about her activities although family lore says that she worked for the Canadian Pacific Railway and appeared on the stage under the name of 'Elsie Kelsey'. Following a report of her later exploits in the Halifax Courier in December 1910 Lizzie made the following observation.

"<u>Puzzle</u> What will she do next? That is the question, Inventor, Waiter, Walker, Actress, Elocutionist etc. Ambition is bred in her, she can't help it. 'What has to be will be' is her motto. She is a born fatalist."

This perhaps sheds some light on her activities. It is also believed that she sang with the New York Opera and we do know she was in the city because there is a record of her marriage on 8th September 1910 to fellow Englishman, Harry Humphries, in Manhattan. It is believed that they met on the vaudeville stage where they had been appearing for several years. The marriage certificate is one of the few official documents on which Harry appears and from this we know that he was born in 1881in London and his parents were George and Anna Catherina. This

The Lady Globetrotter

information has allowed us to trace his family, but they never had a son called Harry, and just to muddy the waters even further, Anna was George's second wife who he married in 1890 when Harry would have been nine years old. There was however a son, Arthur, born in 1881 to George's first wife, Mary, in Hillingdon (not a million miles from London). Was this the elusive Harry and had he changed his name for theatrical purposes? This is possibly him but there is another candidate. George's second wife Anna, who he married after the death of Mary, was herself a widow. She brought two children of her own into the family, Bertha and Warnar Inch (the family were of Dutch origin). Warnar was born in 1881, the same year as Harry, but in Bridport, Dorset. We know Warnar emigrated to the USA as there is a record of his marriage to Amy Kiddell on 20th December 1908 in Chicago, Illinois. After that, just like Harry, the trail goes frustratingly cold.

Whoever the elusive Harry was, the only personal details we have about him come from the interviews he gave to the newspapers whilst he and Lizzie were travelling. He claims to have variously been a soldier in the British army serving in China during the Boxer rebellion in 1900 and in South Africa during the Boer War, although he also claims to have been an engineer and a photographer in that country. He was also a master embroiderer, a skill he picked up while serving on the Indian frontier, and he spoke seven languages. These claims vary from newspaper to newspaper but he consistently tells of two 'facts'. He was a long distance or marathon runner of some repute and he wore a diamond ring given to him by President Kruger of South Africa for winning a 450 mile race in 1898. None of these facts can be confirmed by any official source. Although Lizzie may not have known it at the time, she had tied her fate to an enigma.

Their first act as a married couple was to go on honeymoon, but this was a honeymoon with a difference. They embarked on a walk from New York to Florida and back again, our Harry certainly knew how to treat a girl. There was of course a very good reason for their unusual journey as we will see.

Chapter 2
The Adventure Begins

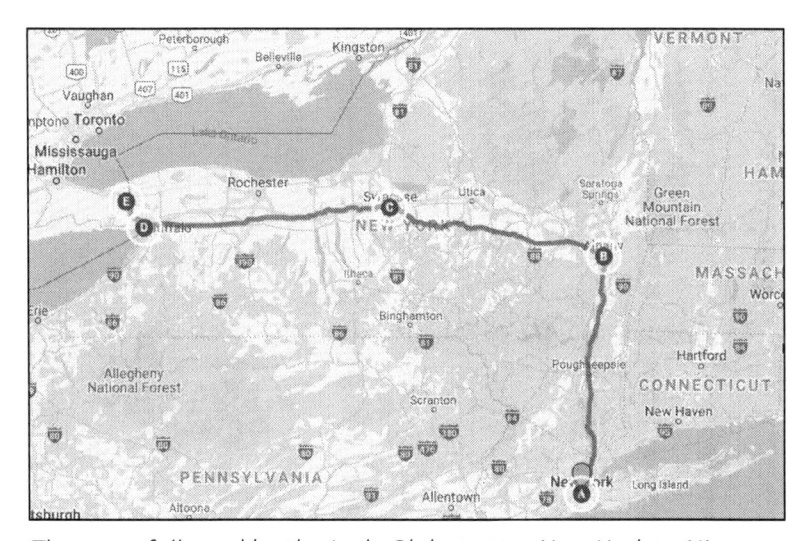

The route followed by the Lady Globetrotter New York to Niagara and then on to Halifax, Nova Scotia
Copyright Google Maps™ Mapping Service

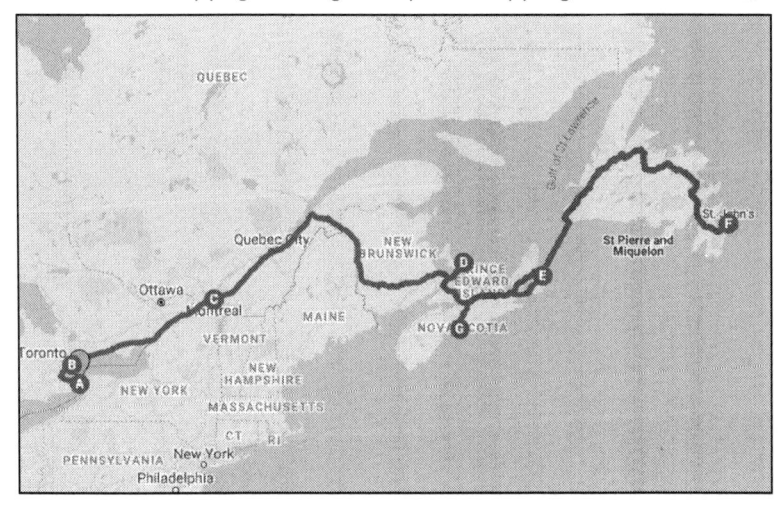

The Lady Globetrotter

Although a 2,500 mile walk would seem like a strange thing to do for a honeymoon, there was method in their madness. It was good publicity for their forthcoming attempt to walk around the world to win a prize of $10,000 being offered by the New York Polo Monthly magazine. How this came about is not clear, had the prize been on offer before their honeymoon jaunt and they were proving their worth; or did the publicity surrounding their extraordinary journey attract the interest of the editor of the magazine and he approached Lizzie with an offer she could not refuse? Lizzie later recounted that she and her husband were in the offices of the magazine discussing with the editor men who had walked round the world, when someone remarked that it had never been accomplished by a woman. She promptly declared that she had a good mind to try it herself, to the eruption of scornful male laughter. This made her more determined than ever to prove them wrong and the wager was made. In a later article in the Halifax Evening Courier Lizzie said that a theatrical manager friend offered the wager on the occasion of their wedding supper, an offer which they heartily accepted.

However the event came about, the intrepid couple set off with great pomp and ceremony on 15th July 1911 from City Hall, New York. The challenge was for Mrs Humphries to circumnavigate the world on foot, travelling 48,000 miles in 4 years so proving a woman's endurance. She was not allowed to use any form of transport except to cross water and a reward of $1,000 was offered to anyone who could prove that she had ridden in any shape or form. They were not permitted to take any money with them and had to finance their journey by lecturing about their adventures at theatres in the places where they stopped en-route. They were allowed to sell postcards of themselves, but were not allowed to accept any other money. Their route would take them northwards to Halifax, Nova Scotia, across the Atlantic to the United Kingdom, across to Scandinavia, southwards through Russia before then turning west into Europe. They would continue south through Africa, through the Middle East into Asia, Australasia across the Pacific to South America and then northwards back to New York.

They were suitably attired in khaki blouses embroidered by Harry with American flags and the words 'Walking 48,000 Miles Around the World', a knapsack with the words 'The Kelsey Kids' embroidered upon it, stout trousers for Harry and a calf length skirt for Lizzie, knee length walking boots with Lizzie's having a one inch heel and, last but not least, an Automatic Savage revolver and holster, no doubt essential accessories in the perilous world of the early 20th century. Their send off is described in their own words.

"We started from the great city of New York, USA one lovely morning in July 1911. Dr Robert Friedman, Secretary of Ensigns presented us with The 'Flag of Peace', the English and American flags with a white border around. It was handed to us by the Hon. William J Gaynor, Mayor of New York, on the city hall steps. Thousands of people gathered there to see us start and to wish us Godspeed on our long and dangerous journey.

After being photographed for all the leading papers we made our way through the crowds to St Pauls Church, there to offer up a little prayer for our welfare and safe return.

Right up Broadway we went, acknowledging good wishes and giving handshakes all the way. On arrival at the tomb of General Grant at 120th Street, Broadway, we were admitted inside by the guard and descended to the vaults. The heavy chains were undone and there in the ice cold vault we laid our peace flag on the tomb as a sign of reverence and respect to the great general's memory. Then, folding up the flag again, we proceeded on our way right through to Yonkers, a good many of our friends accompanying us, and many were the invitations we received all the way up to ride in buggies, wagons and automobiles. Of course we refused. At Yonkers we stayed at the famous Getty House and were highly entertained by several of our friends. We

The Lady Globetrotter

were the chief attraction and had to talk our heads off almost. Anyhow we managed to get to bed about three in the morning. At six we were up again and on our journey."

Promotional postcard showing The Kelsey Kids in their walking outfits

Off they go. The Kids set off from New York

As they journeyed northwards they called in at many places, such as the Remington typewriter works and the Beech Nut Bacon and Peanut Butter factories in Ilion, mainly to allow themselves to be photographed and interviewed by the local papers to gain publicity and also to prove that they were completing the journey as agreed with the Polo magazine. As further proof they often called in at the local town hall or police station and obtained a signed certificate.

They had got as far as Syracuse, only 250 miles into the 48,000 mile journey, when they became footsore and Lizzie had to get an old shoe and cut the top off to enable her to continue, although she was forced to hobble along for a week before recovering. By 11th August the couple were in Niagara where, of course, they donned rubber suits and took the obligatory trip behind the famous Falls.

They were travelling through some very wild and sparsely populated parts of the continent and at one point in Canada they claim to have been attacked by a pack of wolves and were only able to drive them off after shooting and killing one of them with their revolvers. There were other adventures as well as they described in their own words.

"Just outside Toronto we had a narrow escape. Walking very late one night we failed to find a house where we could rest until morning so we went into a little old barn, the only place available. We found some nice clean straw and putting our coats under our heads for pillows we stretched ourselves out to take a nap. In the middle of the night an awful electrical storm arose which set the barn on fire. We woke up in great haste and for a few seconds were almost overcome by smoke. We quite imagined our great 48,000 miles tramp had come to an abrupt conclusion, but fortunately we found the door at the back of the barn where there were a couple of horses and a little calf. One horse was all we could save; the other one we had to shoot. We lost our coats in the flames, the farmhouse was quite a distance off but some instinct must have told the inmates that things were going wrong for someone came hurrying over the fields. We led the horse over to the house and were made welcome. The farmer was glad that we had been able to save something for him and said

that if we had only gone over there before we should have been made comfortable. That was our first real adventure.

We had an extra specially exciting adventure when crossing the river St Lawrence from Cotean Junction to Valleyfield by the railway bridge. This bridge I may tell you is nearly two miles long and as a train might come along at any minute you may guess it was not like walking along the Strand, though that must be a bit dangerous sometimes - at least we thought so when we passed through London. At the end of the bridge there was a guard on duty and he ordered us to go back and get across the bridge by train. That we dare not do however we might have wished to, for if we rode 10 yards we should lose our wager, and our great 48,000 miles walk would be off, and out there in Canada we had attracted so much attention that if anybody had caught us napping, in other words had seen us riding, they would have been after that reward of $1000. So we refused to go back and told that guard the reason why. But he wasn't the least bit impressed; it isn't easy to impress a railway official I suppose. He got quite obstinate and abusive and even swore at us which, considering the fact that a lady was present, was hardly polite was it? But we, prompted by a sudden impulse of self-preservation, combined with resentment of the guard's unnecessary rudeness, seized our revolvers and made as though we intended to fire upon him. Between you and me the pistols were not loaded but how was he to know that? He doubtless took us for a couple of maniacs and he - well, he just funked it. The mere sight of those revolvers was quite enough for him. His bravery evaporated all at once and he shrank back terrified and let us pass. So we got through after all by sheer bluff. For some time, however, we walked backwards so that we could keep an eye upon the angry guard who, seeing that our revolvers were no longer cocked, waxed bold again. But as that boldness expended itself in naughty words we didn't sustain much injury. Of course I knew that we were trespassing and that the *guard was within his rights when he tried to intercept our transit but there, we just set the law at defiance that time anyway and it came off. The wind nearly blew us into the River, it was very exciting I assure you. A train came thundering down and we had only just had time to escape by jumping onto one of the water barriers and holding on tight. That was enough for us. We managed to get to the other end, to Valleyfield. There we were amongst French people who were much amused at our adventure."*

As they had said they had attracted a great deal of attention. In Cornwall, Canada they almost caused a riot when they inadvertently got caught up in a public demonstration against a controversial reciprocity trade treaty between Canada and the USA. They were believed to be spies due to the American flags featured on their blouses and only the swift intervention of a couple of police officers saved the day after they explained that they were the famous 'Globetrotters'. After that nothing was too good for them and Mr Humphries gave a lecture that evening which brought the house down.

Their fame also brought more positive benefits as many people generously offered them accommodation and entertainment including well known figures such as the millionaire George C Boldt, the owner of the Waldorf Astoria Hotel in New York, who invited them to his luxurious Boldt Castle which was located on one of the islands in the Thousand Islands area of New York State. There they met the Spanish ambassador and were royally entertained for the day.

By September they had arrived in Quebec where they were offered free passage to North Sydney, Nova Scotia by Captain Emil Olsen on his ship the SS Wanacousta and from there Lizzie wrote home.

S S "Wacousta"
1911 Sept 23rd

Dear old Dad & Brothers

Here we are on our first trip across the water, on our way home. We are sailing from Quebec, Canada to Sydney, Cape Breton, Nova Scotia, as the Guests of Captain Emil Olsen of Kristiania, Norway, he is a fine fellow & has given us a letter of introduction to his sister & brother in Norway. We are all going to have our pictures taken on the boat in the morning & put it in the magazine. Well we have walked 2,039 miles now from New York, having come through, Albany, Utica, Syracuse, Rochester, Buffalo, Niagara Falls, Toronto, Kingston, Montreal & Quebec. We are now on our way to Newfoundland, Prince Edward Island, New Brunswick, Nova Scotia & then England.

We may be over there in a month from now, it all depends on the weather, it gets very cold here in another month. If you write to the address on the enclosed card "Polo Monthly", 546 Fifth Avenue, New York, USA your letter will be forwarded to us wherever we may be. We are in touch with the office all the time. I shall be glad to see you all again, and shall have so much to talk about. I guess you will look a little bit different than when I left home, & Kelsey must be a big boy now. I will close wishing you good night & good luck.

From your loving children

(Lizzie) Elsie & Harry. The "Kelsey Kids"

"Walking Round the World"

This was the first of several sea journeys they took as they trekked through the islands of north eastern Canada. Once again we will let them tell the tale.

"At Sydney we visited the mayor and the steelworks and walked over to South Sydney where we gave another lecture, and had to wait two days for a boat over to Labrador and Newfoundland. On arrival at Porta-Bass we started out on our long and tedious tramp through Newfoundland. We made fine time, although the walking was so bad. We made our way to Harbour Grace where we took the boat for Labrador called the White Rock. We landed at Hamilton Inlet, after making 350 miles in Labrador, departed the following Wednesday for St John's where we took the SS Kamfjord for Sydney and walked from there through Little narrows, and then, making our way to Pictu, we took the boat to Prince Edward Island.

St John's, Newfoundland has a warm place in our hearts. We received the greatest hospitality and good wishes. Archbishop Howley of The Palace welcomed us warmly and gave us the Holy Seal along with a few words of Scripture which is a great comfort. From Newfoundland to Sydney was an awful trip, the worst known for 60 years. Seven schooners went ashore and were wrecked. All the dishes on board were broken; we were thrown out of our berth and when we got to Sydney we thanked our lucky stars that we were on terra firma again. Through Cape Breton we had hard times, often having to go hungry because the people around were so poor. Of course in a town it was all right but they are about 180 miles apart. We got a good feed of wasters and codfish one day with a couple of old fishermen.

Through Nova Scotia to Pictu we had very severe weather and in New Glasgow I had to stay in bed for a day. Our next place of interest was Prince Edward Island. While there we visited Premier Palmer and Governor Rogers of Charlottetown. On our way to Summerside we had a fine partridge supper given to us by a bunch of young fellows. We enjoyed it immensely and had lots of fun. In Summerside we visited the most expensive Fox Ranch on the island, worth $50,000, then proceeded through New Brunswick to Truro, Nova Scotia, right down to Halifax *where we had our passage* (to London, England) *given to us by Christopher Furness, nephew of Lord Furness, Halifax, Nova Scotia.*

For 13 solid, weary days we were on that boat and the rush we had to get it - only 15 minutes to pack up our things and get on board! It was certainly exciting and I assure you we looked like a couple of lunatics rushing through the town, reporters chasing after us to get a last word. That passage over was very wild and stormy. The first week was terrible and we felt under the weather a little, but managed to get a little enjoyment out of it sometimes. Everybody on board was good and kind and all did their best to entertain us. Captain Chambers was always at our command and took a special delight in making us happy.

We had a glorious time in London and stayed at the Waldorf Hotel where we received special invitations from Oscar Hammerstein to see 'Quo Vadis' at the London Opera House. We had the honour of being escorted by special permission to the Royal Box and sat in the King's Chair."

What Lizzie did not mention in her account was that they were provided with a first class passage with the Furness Line completely free of charge; their growing celebrity status was paying dividends.

Lord Furness started his career as a buyer in his elder brother's wholesale provision merchants. He realised that instead of hiring other people's ships to bring in their stock it would be cheaper to use their own vessels. This was the start of his shipping business which became one of the main employers in Hartlepool, the place of his birth, and over the years owned in excess of 1,000 ships. He was also a Liberal Party politician, serving two terms as the Member of Parliament for Hartlepool. His grandson, also named Christopher, and not to be confused with his nephew, won the Victoria Cross, Great Britain's highest award for gallantry, on 24th May 1940. While serving as a lieutenant in the Welsh Guards in

France he mounted an attack on German forces, delaying them and enabling a transport column to successfully retreat. He was killed in the action during hand to hand fighting.

German born Oscar Hammerstein was the first of the successful Hammerstein musical dynasty. As a young man he ran away from home to America, settling in New York where he worked in a cigar factory. He became a successful cigar manufacturer and invented many machines, mostly involved with the cigar making process. Using his new found wealth he built a number of opera houses and theatres as well as writing several operas. He entertained the globetrotting pair at his own Opera house in London, perhaps Lizzie's own story of leaving home struck a chord with him. Although both his sons continued to work in the theatre business their name is best associated with Oscar Hammerstein II, the world famous American lyricist whose musicals, as half of the partnership Rogers and Hammerstein, include Oklahoma, Carousel, South Pacific, the King and I and the Sound of Music.

The pluck and endeavour of the 'Lady Globetrotter' in her quest seems to have captured the hearts not only of the ordinary man and woman in the street but also those of the rich and influential in society. However not everybody saw their journey as a worthwhile enterprise. The following article appeared in the Pall Mall Gazette of London a few months later when Lizzie arrived in Holyhead, Wales.

Misapplied Endurance

Mrs Harry Humphries of New York is on a walking tour round the world and reached Holyhead a day or two ago en route for Ireland. Her reason for undertaking this itinerary is to answer a challenge and test the endurance of women.

But it would seem to us that the endurance of woman is not being tested at all, but only the walking powers of one woman. And these, we should imagine, were never in question.

The Lady Globetrotter

According to Mr Alflalo, in the new issue of the 'Fortnightly' women have been making journeys since the Queen of Sheba went to the land of Solomon to worry him with riddles, and only the most foolish of the Queens many successes have travelled on foot. Feats of physical endurance, which are seldom undertaken save in a spirit of bravado, imply a wicked waste of energy that might be usefully employed. The old soldier is a noble site, but the old sprinter has only a sporting interest. The popular enthusiasm for muscle training only excites a local and temporary enthusiasm for the muscle strainer. And the strange thing is to find a woman indulging in waste of energy. It is against the best traditions of the economy of the sex, whose highest praise has always lain in the conservation of the precious and in the preservation of health. When Mrs Harry Humphries shall have walked round the world she will have proved nothing save that she wanted to avoid losing a bet - a feeling which is not peculiar to her. Those who know already that in the day to day battle of life women are capable of fine and silent endurance and of necessary mental and physical strain will not feel their convictions any the more firmly fixed. And those who never did think women worth their salt in any capacity will merely record an exception in Mrs Harry Humphries favour as a "well plucked 'un".

It has to be said that the tone and content of this article are not typical of the usual complimentary newspaper and magazine articles which accompanied the Globetrotters journey. Perhaps the editor of the Fortnightly magazine felt the need to demonstrate that there was an alternative opinion to such undertakings.

Chapter 3
England and Home

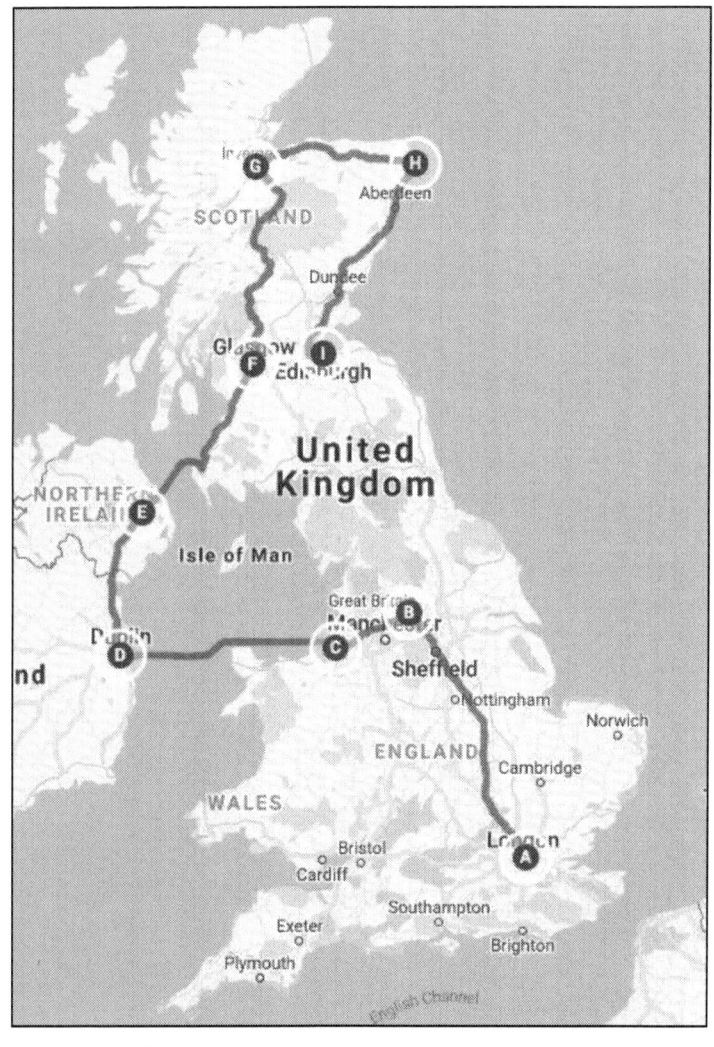

*The Globetrotters Route through the UK.
Copyright Google Maps™ Mapping Service*

On 23rd November 1911 the intrepid pair set off northwards from London to trek through the weather of a British winter. There was a picture of them in the London Daily Dispatch warmly dressed in waterproof clothing. This was however preferable to enduring the extreme cold of a North American winter where they could easily have frozen to death somewhere in the wilderness where shelter and habitation were few and far between. In England accommodation was plentiful and the 27th November found them in Luton where they stayed at the Red Lion Hotel, the proprietor, A W Shearsmith, providing them with proof that they had passed through. Handwritten on headed notepaper he said,

"Mr and Mrs Harry Humphries stayed here one night on their tour round the world and we were very pleased to receive them. Wishing them all success.

A W Shearsmith"

This is a pretty representative example of the certificates they received on their travels but it contrasts with one given to them the very next day by the town clerk at Bedford. Resplendent on Town Hall headed note paper he said,

"I hereby certify that two persons purporting to be Mr and Mrs Humphries attended at the Town Hall, Bedford this day.

Charles Stimson
Town Clerk"

Mr Stimson certainly was not giving anything away was he! It just goes to show that 'jobsworths' are not a modern phenomenon.

As the intrepid pair made their way northwards through the heart of England, Lizzie made observations of the places they visited. She was very impressed with some places.

"We had a real good reception here, sorry to go away. The Mayor was a perfect gentleman and everybody did all they could for us and let me tell you that kindness and hospitality on a trip like this goes a whole long way. We certainly enjoyed ourselves and

intend to visit Northampton again on our return to civilisation",

but had a completely different opinion about others,

"Never want to see or hear of Barnsley again".

Whether this was because they did not like the place or people or because this was where they met an individual who was to figure in their story later is not known.

Whatever their thoughts about their hosts on their journey, they had one objective in mind, to arrive in Halifax in time for Christmas so they could spend the festive season with Lizzie's family. By this time her father and brothers were living at Burnley Road, Sowerby Bridge, her father's engineering company having gone out of business. They duly arrived on 23rd December 1911 to a wonderful welcome as Lizzie tells,

"What a lovely reception we got here on Saturday, December 23rd 1911. Met all kinds of old friends and didn't want to leave. Shall never forget my homecoming. Received lots of presents. Harry met my people for the first time."

So Harry definitely was not the mysterious fiancé mentioned in the photograph earlier in the story. Who was he I wonder? The local paper, the Halifax Courier, covered their arrival in great depth and wrote an article about their adventures. It is worth reproducing in full as it is a typical example of the publicity surrounding the travels and adventures of the 'Kelsey Kids'.

KELSEY KIDS TOUR
ARRIVAL IN HALIFAX
STIRRING ADVENTURES

The 'Kelsey Kids', who are walking around the world, arrived in Halifax yesterday afternoon at 3:15. The Courier readers will already be familiar with them because of the frequent notices which have appeared, their story maybe given briefly. Miss Lizzie Yates, a native of Luddenden, left this district on June 19th 1908 for Winnipeg. After work in a government telephone office and appearing on the stage she was married to Mr Harry Humphries, a professional photographer. They spent their honeymoon making a 2500 mile walk. On their return they entered into a wager with the Polo Magazine Co, New York, for $10,000. The wager is put on Mrs Humphries alone in order to ascertain as to whether a woman is strong enough to stand the different climates and also the continual walking at all times. They have to be back in New York for July 1915. Mrs Humphries at the end of each month writes an account of all the different happenings in the places she has visited, which articles are reproduced in the Polo Magazine. They began their walk from New York in July 22 and with varying experiences they have tramped through Canada, Labrador, Nova Scotia, Newfoundland etc. Lord Furness very generously gave them a free passage across the Atlantic. In London they had a great reception. Many thousands of people thronged to see them so that in seven hours they only walked 3½ miles. They stayed at the Waldorf Hotel. They have now travelled about 5,000 miles.

Mrs Humphries, who is 28 years of age, is only small in stature but she looks remarkably fit. Her attire is designed to give as much freedom as possible in walking. She is an ardent physical culturist, a member of the Health and Strength League and she makes the proud boast that she has never had a headache or indigestion in her life.

Mr Humphries is an Englishman by birth, a native of London, and is 29 years of age. He is of a slender build but continual outdoor life and the walking has made him as tough as pin wire. He went through the Boer War and was in the Boxer Uprising in China. In 1908 he won £1000, also a diamond ring which is valued at about $1500, a 4 carat stone set in a very heavy solid green gold setting, and a medal which he won for a 450 mile run between Johannesburg and Kimberley, the time being 48 hours. He also won a 25 mile race in New York on July 4th last and incidentally defeated Longboat. He has been asked to run at Edinburgh on New Year's Day in the professional 15 miles race for £100 but he has not yet made up his mind on the matter. Edinburgh is somewhat out of their way just now.

Receiving word that this remarkable couple would reach Halifax about 3pm a Courier representative met them at Salterhebble. Attired in khaki waterproof stuff, sealskin boots, and Jaeger wool jackets, they were very attractive. Their jackets concealed embroidered words: 'Kelsey Kids, walking 48,000 miles round the world', but on their sombrero hats the words 'Kelsey Kids' were plain to be seen. Leaving Barnsley yesterday morning at 8:30 they covered 28 miles in 6 hours including a stay for dinner. It was 3:20 when they walked into Halifax General Post Office and not only got many letters, but obtained the towns postmark on their papers. So many friends recognised Mrs Humphries along Commercial Street that it was a considerable time before the Town Hall was reached. The Mayor (Alderman G T Ramsden) graciously invited them into his parlour and while giving them a paper with the towns seal and his signature over it, also that of the town clerk, questioned them interestedly as to their experiences. Mrs Humphries knew Alderman G H Smith and on the Town Hall steps they had a short chat. "Your experiences will begin when you get in Russia," he said, "I have been there."

"Coming through in England," Mr Humphries told the writer, "We have had very good receptions. The people have been very hospitable." "How do you find the roads in this country, as compared with America?" he was asked. "The roads generally are much better here, for in some parts where we have been there were but tracks. Though while in England we have experienced the bad weather we have walked at a record pace. Today's walk has been something of a record and it is very hilly about here."

Nothing can induce this couple to have a ride even were it merely to go back over some of the route. If they are seen riding on anything, the one who witnesses them has but to write to the Polo Monthly in New York to obtain $1000. They have no rest days, not even Sundays. Many might be puzzled as to how they arrange for a change of underclothing but that is a simple matter. In the first place they don't wear much and what little they have is washed overnight and dried ready for the morning. Since they began they have had no illness being only occasionally troubled with foot soreness. On one occasion Mrs Humphries could not get her foot into her boot and she had to hobble along without one boot.

Their experiences have not always been pleasant. Arriving at Belle Ville, Ontario too late to get into proper shelter they slept in a barn. About 2am they woke to find the lightning had set fire to the barn! Mr Humphries, when he and his wife were safe, unfastened cattle and horses thereby saving their lives, for which the farmer was very grateful. Spanning the St Lawrence near a French settlement outside Montréal is an old trestle bridge which has not been crossed for 25 years. The caretaker was going for a gun to warn off the Kelsey Kids, Mr Humphries held out a revolver – they both carry one of the latest pattern such as were used in the Sydney Street siege – and away they went getting over the 1½ mile bridge quite safely. In Newfoundland, when in a dreary waste, 17 wolves came along but the brave couple shot three and escaped while the rest were eating their dead. In

Labrador they had to kill their own game and cook it. Another amusing experience befell them the other day in Nottingham. A farmer with half a dozen cows and a calf was going towards the town when the calf died. The Kelsey Kids volunteered to take the cattle along while he had the calf removed. Away they went, and what a time they had! Those cows would go no way except the wrong one. Still the farmer never came, so after a time they told a cottager to look after them and they left the cows in the highway with her.

Their stay here will probably be until Tuesday, so that Mrs Humphries can visit many old friends and places, including St James School, Luddenden, where on Christmas Day they will take part, still attired in their walking clothing, in some social function.

Their next route will be towards Manchester, Cheshire, North Wales to Holyhead whence they will cross to Ireland. They will go from Dublin to Belfast, Glasgow, Edinburgh, Norway, Sweden, Russia, Germany, Holland, France, Italy, Spain, cross from Gibraltar and go along the West Coast of Africa, along Madagascar, Ceylon, India, China, Japan, Australia, New Zealand, South America, up to Mexico and back to New York. Mr Humphries rattled these off like the reporters call out local stations on the railway, and with as little concern.

"Will you do it?" "We've got to," he answered firmly, and did not seem to doubt but that they would. "Do you like it?" Mrs Humphries was asked. "Oh yes very much," she replied. "We get to see things and places" – and she talked as though she was having a holiday walking tour!

Each carries baggage weighing about 20lbs containing necessaries, souvenir postcards and the like. A peculiarly embroidered flag, the Stars and Stripes, received by them with much ceremony in the USA, they prize very highly. They are taking this with them round the world to exhibit everywhere as *a peace messenger. When they return it will be inscribed and presented to some home of art. An American Journal commenting on them said of Mrs Humphries: "In khaki suit, with khaki satchel held with strap across her shoulders and with high walking shoes she makes an interesting picture, blonde with rosy cheeks and an accent that tells she is an English girl. And she has grit!" We thought she had too as she cheerfully bade us adieu.*

The Sydney Street siege referred to in the article was an incident which occurred in Stepney, East London on 3rd January 1911. On 16th December 1910 a gang of Latvian anarchist emigres broke into to a jeweller's shop in London. They were disturbed by the police and, in the ensuing struggle to arrest them, three unarmed police officers were shot and killed. In the confusion the leader of the gang, George Gardstein, was mortally wounded but escaped with the rest of his colleagues and died the next day. In the following investigation and manhunt two of the gang were traced to a lodging house in Sydney Street, Stepney. 200 police officers surrounded and evacuated the area on 3rd January. When attempting to arrest the men they opened fire wounding a police officer. The police were inadequately armed and on the authority of the Home Secretary, Winston Churchill, a detachment of Scots Guards were summoned from the Tower of London. A stand-off then took place with both sides exchanging shots. Winston Churchill, who as Home Secretary was responsible for law and order, was typically unable to resist the temptation to go to the scene and see matters at first hand, although he denied that he had any role in operational decisions that were made at the time. Shortly before 1pm a fire broke out in the property, one of the occupants was shot dead by a soldier and the other lost his life in the fire. While the Fire Brigade were tackling the blaze a wall collapsed on five of them, one of whom subsequently died from his injuries.

The Lady Globetrotter

With their strict schedule to keep to, the 'Kids' could only spend five days in the company of Lizzie's family, so on 28th December they were on their way again. Lizzie described their departure.

'Goodbye was very hard to say to my father, brothers, relatives and heaps of friends. The streets were lined with people to watch our departure, our hands ached with all the shaking and bidding adieu. But we say to the weeping ones "cheer up the worst is yet to come." And so it is, but we hope not, there is a riddle for you!'

That was indeed a cryptic comment but whatever she meant by it they set off westwards across the Pennines into darkest Lancashire accompanied by Lizzie's father, brother Theo and his fiancée, Ethel Barraclough, arriving in Burnley the same evening where they stopped at the Bull Hotel. At this point the pair parted company with Lizzie's family and continued on to Colne. The first day of the new year 1912 found them in Accrington where their picture was taken and published in the Leeds Mercury newspaper two days later. Accompanying this picture in her scrapbook Lizzie made some unusual comments.

'Shall never forget that day if I live to be 100 years old. Black cats and horse shoes are supposed to be lucky. A large black cat came into our room and I picked up a pair of sugar tongs with a horseshoe on them and had bad luck ever since.'

The next port of call was Manchester where the usual reports were carried by the newspapers, but this time there was a difference – she was alone. Where was Harry? There was a clue in a second newspaper report.

'Mrs Humphries, who started from New York in July 1911, has so far been accompanied on her tramp by her husband, but the

latter has, for the present, given up the task owing to nervous prostration.'

On 5th January Lizzie wrote to her father from the North Western Hotel, Liverpool.

Dear Dad
Just received your letter, Harry's address in Scotland is:
c/o Mrs Geo Humphries
27 Glebe Park
Inniskilling
Fife, Scotland
& in London

c/o Mrs J Colbeck
14 Gloucester Mews East
Dorset Square
London

I am going to see if I can get a train to London and see if he is there, & if not I will go & see Mr Smith, the Editor of Health & Strength & make definite arrangements with him about the Magazine.

Will wire you tonight about Ethel & Theo. I don't know what to do Father. Harry is my Husband after all, if you write to him do not say much or he will not do anything. Only say "why did you leave Lizzie in such a fix" & ask him to be alright with me.

The first address in Scotland is that of Harry's father and the second in London is that of his step sister or sister depending on who we think Harry really is. The reference to Ethel and Theo is probably about their engagement as they were married in the following July. The upshot of the letter however is that Lizzie did not know what had happened to Harry or where he was. She was now in a dilemma about what to do, should she abandon her quest or should she carry on regardless without the support of her husband. She decided, temporarily at least, to continue on to Liverpool but injured her leg somewhere en route and was forced to rest in the city. Her father, no doubt alarmed by the turn of events, hastily set off for the city to meet her and forbade her to continue without the assistance of a man, indicative of the attitude towards women of the time. How could a young lady possibly cope without the help of a male? The Health & Strength magazine, with which she was in regular contact about her travels, was in no doubt as to what she should do. In an article dated 20th January 1912 they rather condescendingly said.

"Those of you who have been following the progress over the Kelsey Kids on their great 48,000 mile tramp will hear with deep regret that Mr Humphries, owing to a serious nervous breakdown, has been compelled to relinquish the enterprise upon which he and his bonny wife set forth with such enthusiasm. Mrs Humphries is pluckily plodding along alone, but she writes me from Scotland that she has also been ill but longs to continue because she cannot bear the thought of giving in. I have strongly advised, however, to do so and I am sure that every 'Health & Strengthite' will agree with me. A tramp round the world unaccompanied is no fit task for a girl – and Mrs Humphries is little more. She would be capable of the endurance required I believe, but there are dangers also that she should not risk. She has done well and if she fails hers will be a splendid failure. These out of the way feats appeal to the imagination but I wonder whether after all they are fraught with any practical value."

As Lizzie wrote to the magazine from Scotland, I can only assume that she had gone to see Harry's family as she was back

The marriage of Ethel Barraclough and Lizzie's brother, Theo Yates, at St James Free Church, Luddenden, Yorkshire in July 1912

in Liverpool by the 18th January where she was joined by Professor William Modley, a 'Physical Culturist' who the couple had met in Barnsley and who joined Lizzie at the invitation of her father. William Modley is an interesting character, he was what we would nowadays describe as a personal trainer. He had previously owned a gymnasium in Barnsley and was actively involved in gymnastics and physical training. He had seen the couple lecture at the Empire Palace, Barnsley where he approached them as fellow fitness enthusiasts, presumably offering his help which was declined. However, with Harry's departure, his services were now required and on 24th January they set off into North Wales. However things did not go well from the very start. On 26th January 1912 Lizzie wrote to her father.

Hotel Metropole
Colwyn Bay
1912 Jan 26th

Dear old Dad

I received your letter yesterday & was very, very sorry to hear you are not well. It seems a shame that I who am trying to do more than anyone else should be the one to cause you so much pain & trouble, but it is something I can't help. I am sure Harry will have to suffer for it in some way or other before he leaves this world, he has made hundreds of dollars through me on this trip & I mean to make hundreds for myself yet.

Poor old Modley is trying to do his best but we are very heavily handicapped in a financial way just at present, none of the theatres here in North Wales are open & we have had to do the best we could. I had to get a block for my cards & order a thousand printing, also had to get real photographs made, altogether we have had a lot of expense getting everything arranged & it has left us practically without any funds in hand. You asked me how we were fixed financially & I am telling you just how we stand.

So far as hotel accommodation goes I am getting put up just the same as usual fine, the best Hotels in every town & everybody tries to do their level best for me, but you see Dad it is from a business standpoint that I am short of working capital. I need £10, to put things in a working condition so that I can go ahead, send Prof. Modley on ahead of me for a day & let him book up some theatres in Ireland & get the Hotel accommodation fixed so that people will be expecting me, & get some bills printed, stickers etc. I know exactly how to make a pile of money on this trip Dad & if you & Theodore would consent to do this for me between you, I will very soon be able to pay you back again. I have got half a day newspapers going with the news already & I do want Modley to go on to Dublin a day ahead of me at least &

make sensible arrangements. If you can at all send me anyhow £5 at once & the other five at your earliest convenience. I promise you to return it before leaving Scotland. You know as well as I do that it needs <u>advance work</u> & everything will be OK. You have heard theatre managers say "if I had only had a day or two" I would have put you on at once. I won't say any more. I know you will help me, I rely on you completely & you will not regret it. What you loaned to Modley has nothing to do with me, that was for <u>his</u> benefit. I am practicing Ju Jitsu as hard as I can, everybody is interested. I have the signature of the Dean of Liverpool Cathedral, Cannon Roberts of St Asaph Cathedral & Countess Dundonald is sending me hers from Gwyrch Castle, Abergele, so you see what people think about me. This is one of the sweetest Hotels in Colwyn Bay.

Please send Postal Order to GPO Bangor then we can start right away & get booked in advance.

Hoping you are all well & happy.

Love from Lizzie.

The Lady Globetrotter in Bangor, Wales. What was naughty about the 'Bangor Bhoys' is not recorded

Lizzie somewhere in Wales

The sudden departure of Harry had obviously left Lizzie in some financial hardship but according to interviews she gave to newspapers at the time, she was hoping to meet back up with him at his parent's home in Scotland from where they could continue the journey together. In the meantime she and Professor Modley struggled on through North Wales and then across to Ireland, often in the most appalling weather conditions.

At the time they were travelling Great Britain was going through several serious crises two of which impacted their visit to Dublin.

The first was the militant Suffragette campaign to gain the vote for women. In 1912 women in Great Britain did not have the vote and there was a growing movement to right this injustice. The great majority of women followed the National Union of Women's Suffrage Societies led by Millicent Fawcett who were conducting a peaceful campaign to win the right for women to vote by legitimate constitutional methods. However the slow progress that was being made led to the formation of the Women's Social and Political Union led by Emmeline and Christabel Pankhurst who not only advocated but carried out violent and destructive acts in a bid to attain their objectives. They became known as the 'Suffragettes'.

The second was the question of Home Rule for Ireland. The majority catholic south of Ireland wanted to become independent from Great Britain but there remained the vexed question of the predominantly protestant Ulster region which desired to remain within the United Kingdom. A proposed Home Rule Bill was causing great agitation within the province just at the time of Lizzie's arrival. Add to this the presence in Belfast of that political giant of the 20th century, and advocate of Home Rule, Winston Churchill, on the very day that Lizzie arrived in the city.

As Professor Modley described.

"We had a very bad time from Dublin to Belfast, it rained nearly the whole time. We were a bit too late for the Churchill meeting, but we saw crowds of people there. We had two tickets sent to us for the meeting, but it was rather too far away when we received them. We had to walk from Dunleer to Portadown (50 miles) in one day, and from Portadown to Belfast on the Thursday (the day of the meeting) and this was about 25 miles. It rained in torrents both days, and we entered Belfast about 8pm looking much like drowned rats. The people crowded round us so much that we had to get police protection. It appears that they had thrown several Suffragettes out of the meeting ground and many people took us to be a new kind of a Suffragette. We had quite a hot time trying to find the hotel, and the police advised us to get in as quick as possible."

Undeterred Lizzie lectured at the Empire Theatre, Belfast before departing by ship for Glasgow where she arrived on Valentine's Day 1912. All this time she had entertained hopes that her husband, Harry, would be meeting her in Scotland after recovering from his 'nervous prostration' to accompany her on the remainder the journey. A letter from professor Modley to the Barnsley Chronicle soon dashed those hopes.

"I am sorry to say that the world walking tour has come to rather an abrupt end. Mr Humphries has sent two letters to his wife saying he will not go on with the trip, and not only that, he has taken all the money back to New York and left her without a penny. I have decided to give the trip up tomorrow, and intend to call and give you the true facts of the case on tour. Mrs Humphries intends trying to do the journey alone but I tell her it is absolute madness because she cannot get along without money. I don't intend going another step without her husband goes too. I am lecturing for Mrs Humphries tonight for the last time and then I leave her and Scotland. Mrs Humphries thinks she is capable of going alone on this trip. She cannot speak anything but English, and her voice is so weak people would not be able to hear her even if they understood the English language. I think she will retire now she is in Scotland and go and live with her husband's mother at Inverkeithing. She has walked about 6,700 miles in six months and two weeks. Her husband is in New York and wants her to give this trip up and go back to him. I think she would do if he would only send her sufficient money to cover the passage."

Lizzie however had different ideas and she wrote defiantly to the Health & Strength magazine.

"Mrs Humphries of Kelsey Kids renown has decided to continue her arduous and hazardous tramp round the world. Her husband, owing to a nervous breakdown, has had to return to New York; but Mrs Humphries, who has the real grit of a Yorkshire lass, is sticking on though. 'I mean to finish,' she says, 'and to show what a girl can do.' As I told you a week or two ago I advised her not to continue and I still adhere to that advice. I think the enterprise is impossible and for an unprotected woman most unwise. But Mrs Humphries' amazing pluck has won my unbounded admiration; I think she is a marvel. I shall keep you posted up with details of her progress from week to week and whether she sticks onto the end, or whether she gives up, I shall cry and you will cry,' Bravo, little woman, Bravo!"

Lizzie in Scotland pictured next to a poster advertising her appearance at a local theatre

She also wrote to her relatives in Halifax after which the following article appeared in the Halifax Courier.

"We are in a position to emphatically deny that the 'Kelsey Kids' walk round the world has been abandoned as was suggested Saturday last by the letter of Prof. Modley who has accompanied Mrs Humphries for a time. Her relatives from this district have this week received a letter from her written from Paisley where, at St George's Theatre, she is appearing twice nightly giving her experiences. She has already been booked for several theatre engagements at Partick. She is very hopeful in her letter about her future and never so much as hints at giving up the walk. Modley, who set out the story that it was to be abandoned, may have been influenced by the fact that his engagement with Mrs Humphries as advance agent had been terminated."

So was Professor Modley sacked as suggested in this article or did he leave of his own accord as he said in his letter. He had promised to tell the Barnsley Chronicle *"enough to startle South Yorkshire a bit"* when he returned but there is no further mention of the Kelsey Kids or his association with them. The paper either decided not to publish his revelations or he simply did not bother to tell them. Perhaps he had his eye on other things as shortly after his return he set out on a tour of his own around Great Britain with his brother giving gymnastics demonstrations. Perhaps he had learned something from the 'Lady Globetrotter' after all.

What is clear however is that Harry was not suffering from any form of 'nervous prostration', he had quite simply abandoned Lizzie and returned to America taking all of their money with him. Lizzie later confided to a Finnish newspaper.

"My husband came with me on the first part of the journey but when he had walked for half a year he got tired of the whole 'Globetrotter' business and took a steamer back to New York. Just like a man!"

Although Harry was now on the other side of the Atlantic this was not the last we were to hear of him.

So Lizzie had once again been let down by another man. First her mystery fiancé, then her husband and lastly her agent. Just to prove that this was a man's world she was barred from crossing the Forth Bridge on foot as women were not allowed on the bridge, even by permit. She did have the last laugh though as she dressed up in men's clothing and sneaked across when the keepers were distracted. Lizzie now took her time in crossing Scotland only arriving in Leith, her point of departure for Norway on 16th May 1912, some three months after her arrival in Glasgow. She was probably taking the chance to replenish her funds by appearing at as many venues as she could before crossing to Europe where her opportunities to lecture would be severely restricted by her inability to speak any foreign language.

Chapter 4
A Scandinavian Excursion

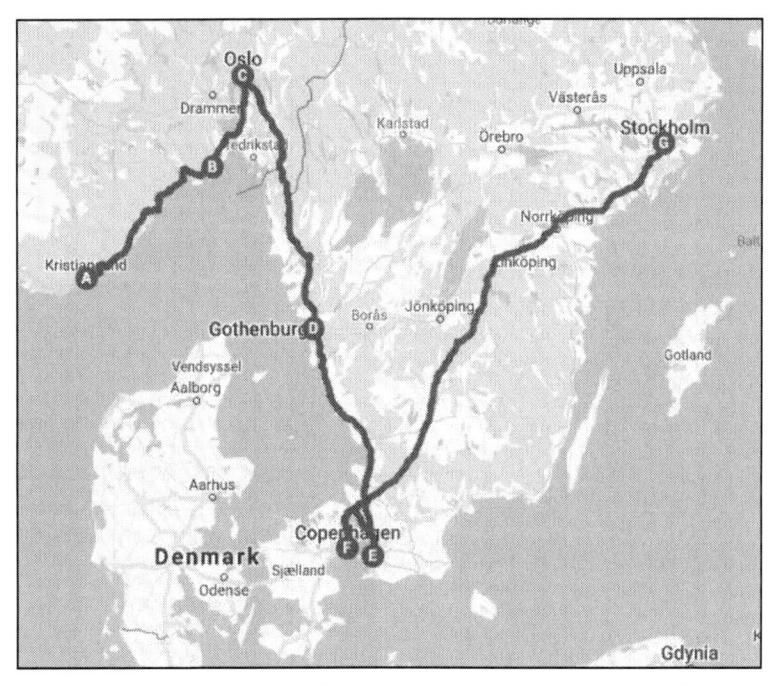

The route through Norway, Sweden and Denmark.
Copyright Google Maps^{TM} Mapping Service

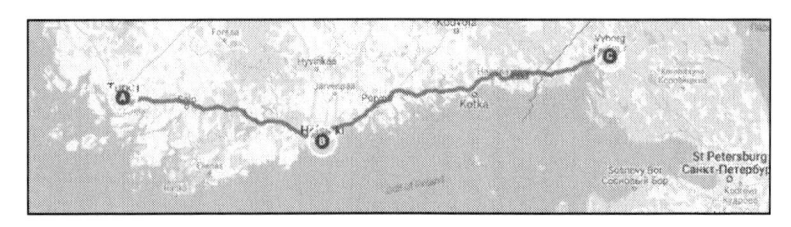

Finland and on to Russia.
Copyright Google Maps^{TM} Mapping Service

Lizzie arrived in Kristiansand, Norway on 18th May 1912. She was now in the land of the Vikings and in a Norway that had only recently become independent from Sweden. She was suitably impressed with the Norwegian people and the breathtaking scenery and after a short break to find her feet, so to speak, she set of northwards along the west coast. At Arendal she found the time to write home.

Hotel Fenix
1912 May 26th
My Dear old Dad

I know you are anxious about me but I have really & truly been so busy that I have not had a minute to call my own, dear beautiful old Norway. I am so glad I had the courage to try my luck out here to show everybody that I am not afraid. I never saw such handsome scenery in the world as I have found here, it is simply <u>splendid</u>. If I had a camera I would be taking pictures every five minutes, every turn in the road brings forth some fresh splendour. The rippling waterfalls, like picturesque cottages nestling under the huge mountains or "Feulds" in Norwegian language, the beautiful fir & pine trees, the "Fjords" & little islands way out in the water, it is a place to rave about, at every house one sees the national flag flying I have walked almost the coast road from Kristiansand and here.

I arrived Kristiansand last Saturday afternoon about 3.30 & went to Olsen's Private Hotel & stayed there until Thursday afternoon because you see being all alone & not speaking the language of the country, not knowing anything of the customs, I thought it the wisest plan to see & know as much as I could before leaving there & I had my picture put in the paper on the Monday as you will see & I also had a lot of bills printed in English & Norwegian because it is very seldom that one can meet a person who does speak English.

The Lady Globetrotter

I had a glorious time in Kristiansand. I met a photographer by the name of Ferdinand Korn & he took eight pictures of me which I am sending to different newspapers, magazines etc. & it all helps but I am afraid I shall find it very hard work to lecture here, the people do not understand the idea of it yet, but really I am progressing very fine indeed to say I have only been here one week. I can understand a great many Norwegian words now & today I met the British Consul of Arendal & his cousin "Margerathia Alligeres" & they are going to take me out in a motor boat down the river tomorrow & then to dinner, & the British Consul told me today that it would give him very great pleasure to pay my bill here at the Hotel for me & that if I needed any assistance to only ask him & he would do anything for me in his power. I tell you this is a delightful country & one of the best I have yet visited, everybody is so polite. The boys in the street all raise their caps almost to the ground when a lady passes by & the little girls curtsey. It makes me feel like a little Queen.

You might write to me c/o Post Restante Drammen, Norway. I must say goodbye now I have such a lot of writing to do & the Spanish Consul asked me last night to go out this afternoon in his motor launch with a party of other gentlemen who have been in New York. I was invited last night to the highest & best gentlemen's' club in Arendal & met the owner of the "Kamfjord" one of the boats I sailed on from Newfoundland to Sydney, Nova Scotia & he told me that in January this year the boat went down & all hands were lost & poor old Captain Salneson one of the best of men went down with her. I have his signature in my book given to me last October.

His wife is here in Arendal & the youngest child is only 11 months old. I am going to see her & try to cheer her up. It seems so strange that I should come here & get to know all about it right where Captain Salneson lived, poor fellow. He gave Harry & I his own room & he laughed at me for being so sick & kept telling me that the ship was going to sink. It was the same boat that we were on when that heavy storm arose last October & we were lost on the water for 2 days but thank God we didn't go down that time but I can imagine the whole scene when she did go down "what has to be will be" no matter how we try to avoid it.

Well good bye & God bless you. I will write again soon.

Your loving daughter

Lizzie xxxx The Lady Globe Trotter

Lizzie's family were indeed anxious about her, her father was moved to write to Lady Ishbel Aberdeen, wife of the Viceroy of Ireland, a keen social reformer and a past president of the Council for Women, an organization advocating human rights for women. Lizzie had actually met her when she passed through Dublin earlier in the year. He received a sympathetic reply with a promise to send Lizzie details of the Council for Women in Norway. Whatever the concerns of her family Lizzie was determined to carry on and by June 7th she had reached the historic town of Tonsberg, reputed to be the oldest settlement in Norway. She related her experiences.

"I shall never forget my first impressions of Kristiania. I used to sit in my room at the Westminster Hotel watching the beautiful girls in fashionable attire strolling up and down the boulevard under the shady trees, whilst the men were lounging on the seats smoking, reading and baring their heads to the breeze. I walked all along the coast from Kristiansaand where I arrived three weeks previously on the 'Thorsa'. How quaint the town looked and how strange the tongue! The first man I spoke to could not pronounce a word of English so I was quite at a loss to understand him. Captain Stark put my baggage in the hands of a Norwegian reporter and told me to follow him to 'Olsen's

Private Hotel'. I did so and was very much relieved to find that Mr Olsen could speak a little English.

I was shown upstairs to my room. I had to cross a kind of courtyard apart from the rest of the house. I climbed two flights of stairs and found myself in a little sunny room looking out onto a garden full of trees in full blossom, and on the other side of the trees was the British Consulate, also the Danish. After arranging my things I paid a visit to Reinhardt, the first British Consul I have had the pleasure of meeting in a foreign country. He signed my papers and then sent me along with a boy from the office to get my roadmaps of Norway from Kristiansaand to Kristiania. It was all very, very strange to me. I had been accustomed to hearing nothing but English and I couldn't follow the girl's conversation in the store where we got the maps. I have since found out that she supplied me with three maps too many for the distance I had to travel in Norway. The boys escorted me back to the hotel and then I had dinner. I enjoyed it very much, some of the dishes being entirely foreign to me. I had some delightful soup which they told me was after the style of Bouillon. It was delicious and for dessert we had 'Raharbra' and cream. In English it means fresh rhubarb and some kind of jelly which mixed together with the rich cream tasted like a new mown hay field smells on a lovely summer's evening. You will think my idea of it very funny but I can't explain it in any other way.

I am very much in love with Norwegian dishes; most of them have a delicious flavour. I was made much of in Kristiansaand and I thoroughly enjoyed myself. I made the acquaintance of Lt Anderson of the Kristiansaand Brigade. He is also a teacher of music. His daughter and I became very friendly and she tried to teach me her native language and very often we both got stuck and had to look at my little phrasebook. I enjoyed my lessons very much and now I can understand a great many words such as

'Var saa god' which means 'if you please' and pronounced 'Varshago'.

There is a curious custom here. When one rises from the table after dining as a guest never forget to say 'Tak for maden' which means 'thanks for supper, or dinner' and is pronounced 'Tock fer motten.' Then again when a person receives any small gift it is customary to say, 'Tusen tak' or 'a thousand thanks' and is pronounced 'toozen tock.'

The roads are very good though, of course not like the old and well worn English ones and the miles are quite different in length. The first Norwegian mile I walked looked almost like a day's march. It takes seven English by land and four by water to make one and if a person walks over here four miles a day it is considered a very great thing indeed. One day I walked five - 35 English miles and at night I felt pretty tired I can tell you.

In Norway the spirit of poetry is everywhere - in the limpid depths of the countless lakes, in the silence of the pine clad forests and in the stately grandeur of the huge mountains. Even as I write these lines I am carried out of myself in spirit to the places I have told you about and my heart is glad to think that God made such splendour for us poor human beings to feast their eyes upon.

Now I have something very sad to tell you. I met at the Fornise Hotel, Arendal between Kristiansaand and Kristiania the owner of a little Norwegian boat I sailed on from St John's, Newfoundland to Sydney, Cape Breton and was horrified to hear that in January this year she went down and all hands were lost. Poor dear Capt Salveson was a brave man and a good one; he left a wife and several children the youngest being only 11 months old. I have Salveson's signature in my book and I feel as if I am in mourning for him.

Dear delightful old Norway, oh how I love it with its beautiful lakes, magnificent mountains and dainty little picturesque *houses nestling under the huge 'Felds' or mountains covered with fir and pine trees. I have had hard times; but, you know, I am a sticker. British grit and American progressiveness combined with the fighting spirit of the Gauls mingling in me for I must tell you that my great grandfather was a Bartholomew and fought in the Battle of Waterloo. So how can I give in unless I were thoroughly disabled?"*

At some point during this period Lizzie visited Christiania (present day Oslo) where she somehow managed to arrange an audience with King Haakon VII. She recounted to a Swedish newspaper.

"I said to the King that I supposed I ought to greet him with a curtsy down to the floor and everyone had to walk backwards to the door without turning their backs to him. He laughed and seemed so undaunting that I dared to send him a telegram on his birthday."

The King must have been suitably impressed and charmed by the Lady Globetrotter as he sent her a reply, although it is not recorded what he said.

Despite the upbeat tone of Lizzie's newspaper interviews and the letters she sent home she was still experiencing financial difficulties, particularly as she was now abroad and her opportunities to lecture, which were her main source of revenue, were limited due to her inability to speak any foreign languages, but she believed she had a solution to the problem. In Denmark she wrote home.

Grand Hotel National
Kobenbaen
1912 June 27th

My Dear old Dad

Thank you ever so much for your great kindness in sending on the cheque for five pounds, it is a great blessing to me just now & by sending it, you have helped me to get one Cinematograph picture. I have had another one taken which I can have for my own as soon as I can send the money for it which will be in English about £5. The firm who took it will keep it for me until I can send the money, & Dad it is a fine picture.

I am dressed in the national costume of Norway & it will be invaluable to me. I can make an awful lot of money with my different Cinematograph pictures when I once get them started. So don't be afraid Dad, anytime you wish to invest, send it to me & the more I get the more I can make. You see the Films do not eat anything, after I once get them, they will go on making money for me all the time, it is the hardest part to get them, but you can always think & know that it was your money which paid for the first one & I am so glad to have it my dear old Dad. Times will alter for us all if you will only take my advice. Don't put your money in the bank or in patents or iron again until you have a lot because the interest is so small that you get. £5 put into a Cinematograph picture, especially of someone who is so well known as I am now, will make at least £20 in one night along with my Lecture, so don't be the least bit afraid Dad. Help me as much as you can for a few weeks & I will amply repay you.

I want to make arrangements to have another picture taken here in Copenhagen, Denmark before I leave, now I have started. I must not miss the principle places, so if you can do so, Cable me another Sovereign or two, so that I can get the picture before I leave. Never mind about having to pay 5/- extra to cable it, you know dear Dad, sometimes we miss the opportunity by delay. I shall *trust* you to do this, remember it is a regular business proposition between us & the money you invest in my picture will be just like you buying a machine for £5 & selling it for £10. The Firm who took my picture in Kristiania think it is a *ripping* idea of mine to have Cinematograph pictures of myself & now good bye Dad.

I shall expect to hear from you by Cable. You know I have been sharpened up in America & I want to show HH that I can make my fortune without him. Cable to this Hotel Sat. or Mon.

Your loving daughter

Lizzie

Cinematograph, or moving pictures, were cutting edge technology of the time and the novelty value of seeing the 'Lady Globetrotter' entering their town or city and pictures of other foreign locations were bound to draw the crowds, making the language barrier a minor issue, particularly if you may be able to see yourself featured on screen if you were lucky enough to be around when the filming took place. Lizzie gives no indication how she came up with this idea but perhaps she was inspired by two English film makers, Mitchell and Kenyon, who toured Great Britain in the early 20th century filming local people going about their day to day business in the towns they visited. The films were then shown in local cinemas and were a great hit as people flocked to the shows hoping to get a glimpse of themselves on the silver screen. The pair came to Halifax in 1902 when Lizzie was living in the area and perhaps she knew of their visit which gave her inspiration. Films at this time were of course silent. The 'Talkies' had yet to make an appearance, so Lizzie commissioned a piece of music from Swedish composer Max Uyma. He composed 'The Globe Trotters March', a stirring composition which would accompany Lizzie as she strode on to the stage to show her films.

Although things seemed to be looking up for Lizzie she was still struggling financially and as can be seen from the previous letter and the following one she wrote from Sweden, she was still dependant on the generosity of her father.

Savoy Hotel
Malmö, Sweden
1912 July 16th

My Dear old Dad

I <u>must</u> tell you Dad that I have had the loveliest Film taken in Copenhagen, Denmark with the money you sent me & it is being shown in one of the Theatres, I enclose the clipping, the Film is one taken of the Zeeland trip with Mr S M Breckwoldt the "Bananeu" & everybody in the city is going to see it. I had to pay Kr80 for it, it would have cost me twice as much only I allowed them to show it in their theatre.

My dear old dad I have made such a reputation for myself in Denmark that when I come back I can make piles of money but I must tell you that with not having a manager of my own I have been cheated out of at least Kr300 on this Zeeland walk not by Mr Breckwoldt oh no but by the manager of the garden where I lectured when we came back to Copenhagen, they promised to give us all the entry money and all we got was Kr29 both together and there were 3,000 people to hear me and the price was Kr1/2 each. It is a dirty shame but never mind I will spoil all their future business for them through the newspapers and they will wish they had thrown the money into the river instead of keeping it but it has made me terribly short again.

I have the chance to get an advance agent on 1st August from Copenhagen. He has not got a lot of money in fact only what will be absolutely necessary but he can speak English, French, German, Norwegian, Danish and Swedish. He is a Dane and lives in Copenhagen, he is acting as valet just now for the Belgian ambassador. I think he's honest and straightforward and will do <u>anything</u> to help me make money and that is what I want.

The Lady Globetrotter

I must have someone to go ahead, I was all right in Scotland when I had my manager and so I will be again when I get another one but not before.

I have got three pictures now, cinematograph I mean, and as soon as possible I will have more, then I can begin my own little show anywhere. I have been trying to arrange a lecture here today but it is the same old story. It should have been arranged beforehand now I am afraid I cannot do anything again. I almost wish I could be real ill for two weeks so that I could go to the hospital and rest until Mr Fossling, the young Dane, is at liberty because I <u>hate</u> to go through all these good towns where I could make some money, it is a great shame.

This is my first night in Sweden, the next foreign country will be Finland, then Russia but I have 420 English miles to walk to Stockholm in Sweden yet. If you can send me money to Malmö wire it. I only wish you could see the cinematograph picture of me in Copenhagen it is a regular beauty, everybody likes it, it is very funny to sit in a theatre and see yourself laughing and talking on the screen.

I can hardly realise that Theodore and Ethel are married, it seems so strange. I hope they are <u>happy.</u> I had a postcard from H yesterday he said he was taking <u>a little trip out to San Francisco</u>. I guess he has money enough to enjoy himself even if his wife is starving. He said that you had not answered his letter, nor had I, and that he guessed the whole family was <u>sore</u>, well let him think so it will do him good anyhow. I am meeting the finest and wealthiest people in every country and I do hope that some of them will put their hands in their pockets and say "here you are, take this to keep you on your trip" but I have to wait for the opportunity because if I asked them myself then I lose their friendship and it may be valuable someday. Now I must close. I saw Capt Stack again yesterday, he is a fine fellow. Good night it is 12:30 and I have a very hard day in front of me tomorrow.

Your loving daughter
Lizzie xxxxxxxx
I have got another medal now from Copenhagen, Denmark and I am so proud to wear it.

Once again a request for money and a lament that another man had cheated her. The necessity for an advance agent was obvious but Mr Fossling the Dane mentioned in the letter does not appear in any future correspondence so he probably wasn't hired or did not last very long if he was. As for the other people mentioned in the letters I have been unable to find out anything about them except of course for HH or Harry Humphries, the errant husband. He did take his trip to San Francisco where he fell foul of the law as this article from a San Francisco newspaper shows.

WARRANT ENTERS RACE WITH MARATHON RUNNER
Wife of Harry Humphries Accuses Him of Omitting to Provide

Charging omitting to provide for his minor child, a warrant was issued yesterday by Police Judge Shortall against Harry Humphries, who is said to be a professional marathon runner.

His wife, Mrs May Humphries of 1762 Market Street, is the complainant and alleges that her husband deserted her, taking $700 of community property in money. She believes he has gone to Los Angeles and suspects an affinity. She accuses her husband of omitting to provide for their minor child, Arthur Humphries, aged three years. Mrs Humphries said her husband held world records as a marathon runner.

Although this is not conclusive proof that this is the same Harry Humphries it would it be an amazing coincidence if there were two men in San Francisco at that time with the same name claiming to be professional marathon runners.

So Harry had another wife he had run out on taking their money, and a son, Arthur. Perhaps a clue to Harry's real identity. Was the son named after him?

In July 1912 another article appeared in a Los Angeles newspaper.

HAMMERSTEIN WAGER AMOUNTS TO $10,000

Oscar Hammerstein, of worldwide theatrical fame, hasn't been in the United States very long after his recent return from London until he had wagered $10,000 with Mr and Mrs Harry Humphries on a 65,000 mile motorcycle tour around the world, and now the riders have left their home in Los Angeles to win the $10,000 and encircle the globe astride the two wheeler.

The schedule provides for an average of 65 miles to be covered daily and they expect to spend three years en route.

Mr Humphries is accustomed to covering long distances. In 1898 he won the South African marathon, covering 450 miles in 48 hours running time. He and Mrs Humphries will carry a pennant of the Angel City Motorcycle Club of Los Angeles as a "passport" in foreign motorcycle circles.

The mention of the South African marathon definitely proves that this is Lizzie's husband. This alleged feat appeared time after time in their interviews with the newspapers during their travels from New York until Harry left in England. May Humphries said she believed Harry had gone to Los Angeles providing further evidence that the two Harrys are one and the same person. This begs the question who was the Mrs Humphries who was to accompany Harry on his new venture? It is unlikely that May, who he had previously abandoned leaving her penniless and who was chasing him for the missing money, would choose to be involved with him again and Lizzie was definitely in Europe at the time. There is a postcard in the collection of a smiling Harry astride a Minneapolis Two Speed motorcycle with 'Mrs Humphries' sat jauntily in the sidecar. The caption reads,

Mr and Mrs Harry Humphries, round the world on a Minneapolis two speed motorcycle using Goodyear tires. Started from Los Angeles, California. Finish Panama expected 1915. Total of 65,000 miles. Mr Humphries holds two world records made in New York 1911 and the International in South Africa.

The 'wife' is not Lizzie, but there is a clue to her identity in an article that appeared in the Joplin Morning Tribune in April 1913.

WILL NOT RETURN TO HER HUSBAND

She Fled with Marathon Runner – Husband Offers to Pay Her Fare Home

Although her husband William P Fleming of Los Angeles has offered to pay the expenses of her to return to him, Mrs Fleming, who eloped to St Louis with Harry Humphries, a marathon runner, declares she will not go back.

Although she has been living with Humphries at 2639 Washington Avenue under the name of Mrs Humphries, she declares that her relations with the man have only been platonic. Humphrey is held for the Los Angeles Police, who have a wife abandonment warrant for him, sworn out by his wife in Los Angeles.

"I left college when I was 13 years old to marry my husband," she said. "We have been married 11 years. He is 43 years old and I am 24. My life was miserable with him. I fear now to go back for fear he might harm me. I met Harry several months ago. He did not get along well with his wife, and when she ordered him to leave, he decided to do so. I also had decided to leave my husband. Harry and I started across a desert in our automobile. It got stalled in the sand and we had to give it up. Both of us are fond of literary pursuits and we decided to come east and gather material for book and magazine articles."

Surely this is the woman with whom May Humphries believed Harry was having an affinity in Los Angeles. How easy it would be for a congenital liar like Harry to pass her off as his wife. In any event their escapade came to a sticky end as did Harry's jaunt with Lizzie as one of the 'Kelsey Kids'. Did Harry abandon her as he did with both of his wives? We shall see.

In the meantime Lizzie continued her journey northwards stopping at towns and cities en-route where, now armed with her

The Lady Globetrotter

moving pictures, she gave lectures on her travels. Early in September she arrived in Stockholm where she stayed for several days before taking a steamer to Helsinki, the capital of Finland. At that time Finland was an autonomous Grand Duchy of the Russian Empire and had been so since 1809 when it was annexed after the Finnish war between Russia and Sweden.

She continued westwards on to Vyborg, which at that point in history was on the Finnish side of the Border with Russia. It was there that she met with George Easton, probably through her connection with the Health & Strength League. He was a Scot working in the country in the timber industry and is widely credited with introducing Association Football and curling into the country. In 1908, while playing for Unitas Sports Club, he was part of the team which won the first Finnish football championship. He was also an outstanding athlete as he modestly told Lizzie on the postcard he gave to her as a memento.

"Wishing you good luck. George W Easton, Vyborg, Finland.

First to reduce Finnish record of 100 meters to 10 4/5 seconds.

Times for 150, 200 and 250 meters are 17, 23 1/5 and 29 sec respectively, all under Finnish records still unbeaten. Total abstainer from smoking and a follower of 'Health & Strength'. The pioneer of football in Finland."

It was at Vyborg that she crossed into mysterious Russia with her permit to carry her revolver and ammunition whilst in the country. She had also acquired a companion for her lonely travels, a dog called Vicksie, a gift from a Finnish seaman. Vicksie was to accompany her for the remainder of her adventure.

The Lady Globetrotter

The Lady Globetrotter, taken in Finland.

Chapter 5 The Land of the Tsars

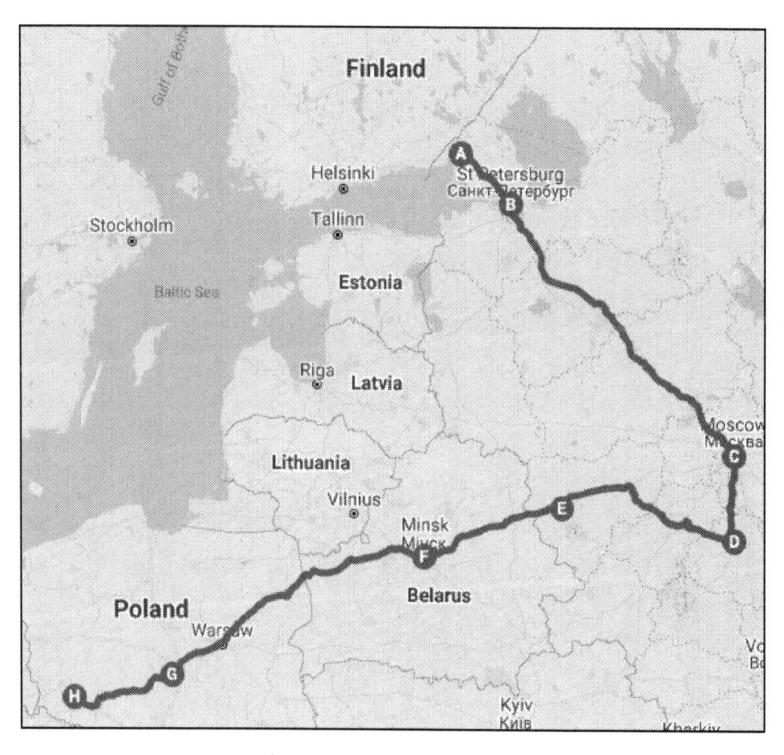

Through the Russian Empire
Copyright Google Maps™ Mapping Service

In 1939 that master wordsmith, Winston Churchill, described Soviet Russia as

"A riddle, wrapped in a mystery, inside an enigma."

Although this was said some 27 years after the time of Lizzie's adventures, it could just as easily be said of the Russian Empire under the rule of Tsar Nicholas II. Their unit of measurement was the 'verst' (approximately $^2/_3$ of a mile) and they used the Julian calendar which was 13 days in front of the Gregorian calendar used by the rest of the world. They even used a different gauge on their railway lines to the rest of Europe to hamper the movement of an invading army such was their paranoia about invasion by a foreign power

By 1912 The Romanov family had been ruling Russia for almost three centuries and Nicholas was the last of the autocratic rulers in Western Europe. He was also the last of his line although he would not have known it at the time. In spite of attempts to modernise their society, Russia was still very backward with few modern roads and a very large illiterate peasant population who had lingering memories of the slavery of serfdom which had only been abolished 50 years earlier. Bad living and working conditions, high taxes and a lack of democracy led to a society seething with discontent. There had been an attempted revolution in 1905 which led to the formation of a Duma, or parliament, although the Tsar took very little notice of it and his iron grip on the country was barely eased.

It was into this mysterious nation that Lizzie now set forth on the next leg of her quest. Just to add to the mystery, no correspondence to or from Lizzie during her time in Russia has survived. It is unlikely she did not bother to keep in touch with her family so any letters or postcards she sent have either been lost or were swallowed up by an unreliable postal system. The only record of her time in the country comes from some newspaper articles and certificates she kept in her scrapbook and from them we are able to track at least some of her journey. Tellingly there are none of the notes and observations Lizzie usually made in her scrapbook against the various articles, perhaps a clue to the trials and tribulations of this part of her journey.

She arrived in St Petersburg on 25th September 1912 and remained in the capital until 2nd October giving lectures about her adventures and raising funds for the remainder of her travels.

Forsaking the roads, which by this time of year would have been impassable seas of mud, she followed the railway to Moscow stopping at a number of places on or near railway stations on the way. At a stop called Okulovka, some 150 miles from St Petersburg, she related an unpleasant meeting she had there with a group of aggressive sounding men who demanded that she gave them her dog. She naturally refused and took out the revolver that she had with her and threatened the men with it, which prompted "the hooligans to flee". Probably as a result of this encounter she spent the night at the home of a local policeman. She finally arrived in Moscow on 9th November having walked 400 miles from St Petersburg. She was now travelling into the teeth of the harsh and unforgiving Russian winter, the same winter which had defeated Napoleon in 1812 and was to defeat Hitler in 1941. Temperatures often fell to -30°C with heavy snowfall and it is not surprising that Lizzie became ill and was advised by a doctor that her exertions were placing a strain on her heart. In spite of this warning she continued to give lectures with the help of an interpreter, N M Blumenfield, and she appeared at the Petrovsky Theatre in Moscow between the 23rd and 25th November. It is not recorded when she left Moscow but on 8th December she was a guest of an Englishman, Edward Charnock who worked at a cotton factory in Serpukhov, 68 miles south of the city. She continued southwards arriving at Tula on 14th December. While in the city she gave more talks about her travels again with the help of Mr Blumenfield, who had presumably followed her, and she visited Tolstoy's grave. It was then that she turned westwards following the railway to Kaluga where she arrived on the first day of the New Year 1913. Later in July 1913, when she was in Leipzig, Germany, Lizzie sent a postcard to a friend in New York showing a picture of herself, her doctor and his brother taken near Kaluga where she says she had a severe illness of the throat. Other newspaper articles indicated that she had been ill on a number of other occasions, which is not surprising seeing as she was battling through rain, mud, snow and freezing temperatures.

Lizzie pictured with a group of unidentified people taken in Kaliska, Russia

After recuperating she continued westwards along the railway line to Smolensk which she entered on 12th February 1913 and on to Brest on the 5th April. She entered Warsaw, the Polish capital on 8th May where she stayed for around 10 days before continuing to Lodz and Pabianice. Poland had a long history as a sovereign nation but that had all ended in 1798 when it was occupied by the Prussian, Austro-Hungarian and Russian empires who all took a share of the country thus ending hundreds of years of independence. The Poles were, and are, a fiercely patriotic people and there were several unsuccessful insurrections against foreign control. Lizzie, however, was oblivious to the rich history of the country as in Pabianice she received tragic news as she recounted in this letter.

Theatre, "Luna"
Pabianice
1913 June 13th Friday

My Darling Dad & Kelsey

My heart is almost turned to stone, my poor brother, oh why is God so cruel, for what reason has he taken away a boy who was just beginning life & the only one you had at home to whom you could confide, I am not forgetting dear little Kelsey when I say this but he is too young yet to know about any business & Oswald <u>could</u> understand. Father I almost feel like a murderer, because I feel that if I had said <u>yes</u> to Harrison's telegram instead of <u>no</u>, everything would have been well, but God knows best maybe.

My poor old Dad what will you do, <u>try</u> to keep up for the sake of Kelsey & don't mind about <u>me</u> you poor dear man, has God no pity at all. I am sure you do not deserve all the bad things that are coming your way, oh, it is <u>too cruel</u> to think of & Oswald was the only one I did not say "good bye" to when I was home last, you remember Dad, you & I walked over the hills to Aunt Sarah Jane's and when we came home Oswald had gone to work and, I have never seen him since & now it is too late. He has been so anxious to go with me, he sent me a confidential letter a few months ago saying he would like to join me in Paris & failing that, he would like to go to America. I sent him a long letter in return giving him my advice but now he has gone to a country where everything will be alright for him, this afternoon I thought my heart would burst with pain, in mind I was with you all at Pellon & by my Mother's grave but I cannot express my feelings on paper, oh why am I so far away from you now at this time of need.

I hope Ellen is with you she is good & will comfort you but, oh my Dad, it is almost too much to bear. I have to go on the Stage

The Lady Globetrotter

in a few moments & laugh & talk & appear jolly, when my heart is as heavy as lead & I know that your home is desolate tonight. I have just sent you a cable of sympathy. I hope you will not be cross. I <u>must</u> do it even if I had to starve to pay for it, which thank God I don't have to do, though things are bad enough here also, but never mind. I could not let you lock the door tonight & have no word from me, in mind I am not with the people here, but at home with you & Oswald, his spirit is free now. In life I refused him to go with me (as I thought for the best) but now I shall welcome his spirit & maybe God will appoint him my Guide for the rest of my trip, anyhow it is a comforting thought to me to feel that he knows & sees my life as it is, because Dad I am sure that what he wanted to do so much in this world & could not, God will allow him now, in some way we do not understand, he was a good boy & so I am not afraid for the suddenness of his death. He had nothing to repent & his unconsciousness was a blessing to <u>him</u> however painful it was to you, Dad. I can well understand how awful you all felt though when you could only watch death creeping on & not speak to him, but tell me if it will not pain you too much, everything which happened, your letter was very short & I can understand it because you <u>could</u> not do otherwise, & now I cannot write more. My heart is aching too much. God bless & keep you till we meet again is the prayer of your wandering daughter.

Lizzie xxxxx

If I could only be with you & comfort you how much better I should feel. It is terrible agony to be here amongst foreigners & to know that one is beyond recall & my other dear ones are suffering the pangs of parting. God help us all.

This heartfelt outpouring of grief was prompted by the news of the death of her 21 year old brother Oswald in a motorcycle accident. As was the case with many young men of the time, Oswald was fascinated with the new science of the internal

The Lady Globetrotter

combustion engine, very much like young people are interested in the digital technology of these days. As an engineer he was, in the words of his father at the inquest, "a very clever young man at anything connected with mechanics", and repaired motorcycles as a hobby. On 2nd June 1913 he was road testing a motorcycle he had just been working on at Burnley Road, Sowerby Bridge when he lost control of the vehicle on a bend, struck a stone which was lying in the curb and hit his head on a telegraph pole which fractured his skull and rendered him unconscious.

As was usual with anything to do with Lizzie's story, the subsequent events were not straightforward. A nearby doctor was summoned who immediately rang for the Sowerby Bridge ambulance but it was unfortunately already attending an incident and was unable to respond. Undeterred the doctor then rang for the Halifax ambulance and was brusquely told they were unable to come as the incident was not within the Halifax Borough and Oswald was forced to wait until the Sowerby Bridge ambulance was available before he could be taken to hospital in Halifax. The coroner asked why nobody had thought to move Oswald to the other side of the road which would have placed him within the Halifax Borough to which the doctor replied, 'you do not always think of those things'.

Two days after admittance to hospital Oswald suffered a series of fits and was operated upon on 7th June but unfortunately died the same day. He is buried in the family grave at Christ Church, Pellon.

The Halifax Courier reported on the inquest under the following dramatic headline.

Motor Cyclist's Fate
Burnley Road Tragedy
Searching Questions at the Inquest

They concentrated very much upon the apparent indifference of the Halifax ambulance staff to Oswald's plight and to the presence of the stone at the side of the road which was struck by Oswald's motorcycle and which was presumed to have been placed there by local residents to enable them to cross into the road when there was water running along the gutter. It turned out that the Halifax ambulance was not allowed to attend incidents outside of the borough except in very special circumstances and no further action was taken or recommended about the actions of their staff and there seems to have been no follow up regarding the errant stone.

Although Lizzie was undoubtedly grief stricken it was simply not possible for her to return to Halifax. This was before the days of airline travel and the only means that she would have been able to use to return to England would have been by rail and she would undoubtedly have returned too late to attend Oswald's funeral. She continued on but it took her one month to travel the 150 miles from Pabianice in Poland, across the border to Liegnitz in Germany. From there she wrote home.

Hotel Reichshof
Liegnitz
1913 July 13th

My Dear old Dad

Many thanks for the money you sent me to Breslau. I could not have done without it & I shall try my very best not to ask you again for any, but oh, I am having it hard. Mr Litwak has gone away. He robbed me many times but he does not know that I am aware of the fact the last Hotel bill in Kalish he said was 60 roubles & it was only 44 & 60 copecks & he told me that he got only 2 roubles discount & he got 10. O, my dear old Dad, there are such rogues in this world, also Mr Litwak has taken my alarm clock & a little leather bag & I have missed a beautiful large photograph of a famous man in Poland. His Excellency General G A Selon & I am sure that Mr Litwak has taken it because <u>no one else</u> even touched my baggage & he helped me to

The Lady Globetrotter

pack up my things so that the Officials at the Frontier would not trouble to look at everything & I am very angry with him. I told him I did not want his help & now I see why he wanted to do it, in order to get the photograph & I am afraid to write to General Scalon & tell him what I think because Litwak told me that he is a member of the <u>Russian Secret Police</u> & if that is true he has <u>terrible</u> power over <u>anyone</u> he wishes to injure for purposes of his own. He could get hold of my baggage somewhere & insert papers which would show that in Russia I had been <u>spying</u> on the <u>Government</u>, then he would give information to other secret police & I would be <u>arrested</u> under suspicion, then the papers would be found in my baggage & I would be sent to <u>Siberia</u> for life, <u>that</u> is the way he would have his revenge.

My Dear Dad, although I have got through Russia so safely, I have not been <u>blind</u> to many things which go on there, but I have been <u>careful</u>, diplomatic & <u>simple minded</u> & so I have got through & so I must be in every country & no harm will come to me, barring accidents.

I depended so much on Litwak because he told me he was a "Freemason", but I don't believe it now or if <u>he is</u> & it were known what kind of man he is, he would be at once <u>expelled</u>. Now I will say good bye, I hope you & Kelsey are well, write me to Dresden Poste Restante.

Your loving daughter

Lizzie, The "Globe Trotter" xxxx

Although no other correspondence has survived from this time it is clear that there must have been some as Lizzie had requested, yet again, money from her father to allow her to continue her travels. As usual she had a tale to tell with the sinister Mr Litwak, who presumably was her agent and guide in Russia, having defrauded her of money and who also turned out to be a member of the Russian secret police, the dreaded Okhrana. If Lizzie was correct she was quite right to believe that she could have been in danger as arbitrary arrest, detention and even torture were a feature of the activities of that organisation. Whatever the case she was now safely across the border in Germany.

The Lady Globetrotter pictured with some very official looking people at Minsk Railway Station, Belarus

Chapter 6
An Outrageous Stunt and an Old Flame Returns

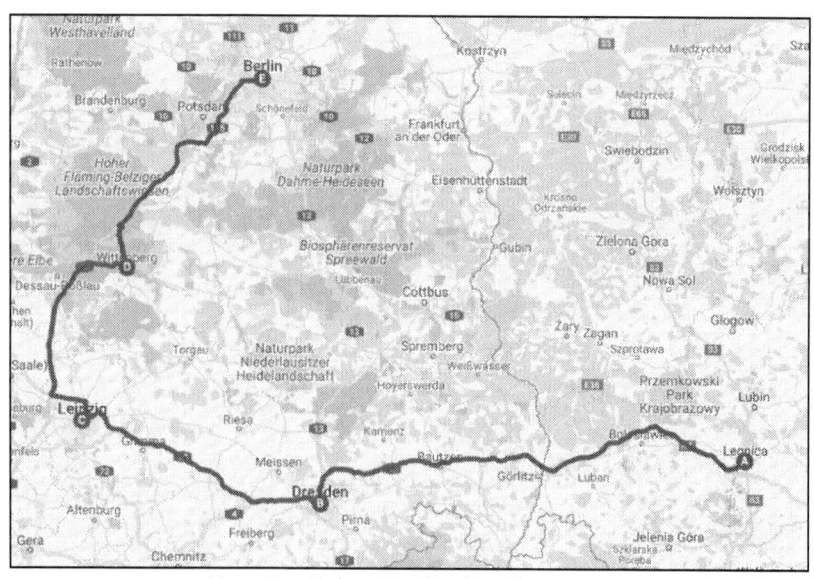

Liegnitz, Poland to Berlin, Germany
Copyright Google Maps™ Mapping Service

A week later, still in the same hotel, she wrote an intriguing letter home.

Hotel Reichshof
Liegnitz*1913*
July 20th
My Dear old Dad
I have had such nice letters from Percy, oh I hope that sometime we shall be <u>really & truly happy</u>, he still loves me just as much

as he used to & I think we both made mistakes, you know Dad I <u>never</u> really <u>forgot</u> about him though I married in New York or <u>thought</u> I did, but it looks as though it is <u>ordained</u> that I must have no <u>real</u> Husband only Percy after all & still everything which has happened is part of a <u>scheme</u> which God has planned for some purpose.

You remember how I used to write poetry & stories & how you tore them up when you could find them, well Dad you can <u>never</u> kill what is in a person's makeup, it is impossible. I am fairly on my way to write such startling stuff as it is the lot of few people to do. How very, very strange is life & I wonder how <u>yours</u> will turn out yet. Don't think I have forgotten about poor Oswald because I do not say anything in my letters, it is because he is <u>always here with me</u>. I do not feel that he is <u>dead</u> but much more alive than before. I am satisfied that he can go with me in this way & Dad don't forget that <u>he</u> has brought back to me <u>help, life & love</u>. It is by his death that <u>Percy & I & you</u> have been brought together again so forcibly because although I had the letter from him in Moscow & again in Lodz I never intended to listen to words of love again from him, he wrote me that he would be my "sincere friend" as you know & I had made up my mind <u>never</u> to marry again, least of all to <u>him</u>. I never thought for one moment that he would ask me again but you two men could not meet each other & put things on a <u>friendly</u> footing after what each knew happened in the past so you <u>sealed my future for me</u> all <u>unknowingly</u>.

My dear Dad I can talk to you now as I never <u>dared</u> to when I was at home, I feel as if your spirit & <u>mine</u> are in the same mold, restless, ambitious, progressive, you will make a <u>good</u> <u>Husband</u> so far as supplying all the wants & comforts of home etc. but there is a <u>something</u> in your <u>makeup</u> that is akin to mine, something which you cannot control which takes you out of yourself away & apart from every human being, something which makes you <u>glory</u> in your heart solitude & you <u>withdraw</u>

yourself into yourself & find in unknown regions heaps of comfort which no one else can impart to you a kind of *spiritual* comfort & elevation.

I *know* that you feel like this although maybe you will laugh heartily over my words, but at the same time you will admit that it *is* so & *I* have that feeling so strong that I am almost *afraid* of ordinary married life though I would be surrounded by every comfort & loved devotedly. I feel as if sometimes I should want even to be free from my body & fly in realms of space. My dear Dad can you account for it? I would dearly love to find out from what *stock* we spring on each side on both the Kelsey & the Yates side.

I know very little of my ancestry. If you will write to "Somerset House" London & enclose one shilling you will receive a correct family tree report. *Please* do so & send it to me. I think that on *your* side all are French but from Artistes or cobblers or what I don't know. Now good bye. *I* feel like a "Joan of Arc" or "Napoleon" myself & I think you made me out of "Cast Iron".

Lizzie xxxx

From the contents and tone of this letter we can only conclude that 'Percy' must be the mysterious fiancé who appeared in the crumpled photo mentioned at the beginning of our story. The letters Lizzie previously received from him have not survived so we do not know what he said but it seemed that he and Lizzie had rekindled their relationship to the extent that he had again asked her to marry him. She had obviously decided that every cloud had a silver lining, and Oswald's death had brought Percy, her father and herself back together and allowed them to put aside whatever previous differences they had. So things appeared to be looking up for Lizzie and, as was her habit, she visited the offices of the local newspapers in Liegnitz and then in her next destination, Gorlitz, to announce her arrival and the dates and times she was appearing in the local theatre. However only three days later things were not looking so rosy.

Hotel "Vier Jahreszeiten"
Görlitz
1913 July 23rd

My darling Dad

I have just received your letter in the Post Office here & while I am glad to have it I am very much upset also because I <u>cannot</u> give up my trip in Paris. I understand you quite well, no I cannot expect you to keep on helping me, oh it is a shame that you have had such hard luck in life yourself. I wonder why it is, I think it is because you <u>trust</u> people too much & you are too good natured like myself. You & I can hardly say "<u>No</u>" to anybody & it is so bad for ourselves. Oh, if you were only here & I could talk to you.

If Percy does not want to help me I can't help it, he has told me that he has <u>plenty of money</u>, nowadays but that it is tied up in shares. He has asked me to wait for a month or so & then he will let me have a <u>lump sum</u>. Of course, it is none of my business what he has told you but Dad as long as I live I shall <u>never be satisfied</u> if I give it up, so long as I have strength to go on, & I know that I could <u>make money fast</u> if I had 50 English pounds in hand. What are these Theatre managers doing when I am appearing in their theatres, making money of course & many times they have announced in the papers that I have rented the theatre from them for 1 or 2 or 3 days, so as to make people believe that I am getting all the money & in fact <u>I</u> am getting very little from them.

In Russia one theatre made in 3 days, 800 roubles, 1,600 marks or 75 English pounds while I got 45 roubles, less than £5. They <u>wanted</u> to rent me the theatre at first for 30 roubles per night but I could not take it & they were very glad after that I <u>didn't</u>

The Lady Globetrotter

but never mind Dad I will find some other way to accomplish my desire. In the end <u>you</u> will benefit by it.

I do not say very much in thanks for all you have done for me but I will <u>show</u> you as soon as I can. You shall not have to work hard in your old days & still you will not be dependent on a wife or a son-in-law, you understand me? When you were near to failing in business you remember how I tried in my girlish way to save you by going to see Greenwood in the Crags, well I could not do it that time but <u>this</u> time I <u>will</u> or else go under myself. If I give up now I can <u>never</u> do it & as for writing a book, I can only do that if I <u>finish</u> it, otherwise everything shall be <u>buried</u> with me. I have suffered too much to <u>hide</u> everything & write a <u>false</u> story & a <u>true</u> one can only be written providing I reach New York.

So my dear old Dad, you will not say again "give it up" please, & now I will tell you how I have arranged things in this town.

I went to see the Director of the largest Cinematograph Theatre The "Union" last night. Well at first he said "<u>no</u> it is <u>summer</u> time & I have all the pictures I want & I have no time to spare for your act". Then I said to him "Now look here did you ever have anybody in your theatre who was willing to give a part of their earnings to a charitable institution in this town?" & he said "no, they keep it all themselves" & then moved like to ask me for more. Well I said "If you will allow me to sell my photographs in your theatre after my pictures (Cinematograph) have been shown, I will give <u>half the money</u> I get to the "Martha Heim" (that is a place for poor girl orphans). He said "Good", you have a heart in you, I will give you 30 marks a night for showing your pictures, Friday & Saturday so I shall get 60 for my work & I <u>should</u> make 100 by my cards, 50 of which I must give away but you see Dad people will give me more money than usual for my cards when they know that part of it goes for the poor people of their own town. So you see I am doing some good *also & it makes my heart glad & I don't lose anything by it in any way.*

In the last town in Liegnitz I did the same. I gave to the "Martha Heim" 23 marks. I made 46 & how glad they were & I have their prayers all through my journey. The Theatre Director in Liegnitz did not want me to sell my post cards at all he said it lowered the tone of his theatre & then God said to me quickly "tell him you will give <u>half</u> of it to the poor people in the town" & the man was surprised that a perfect stranger should do so & he afterwards said to someone "it will teach <u>our rich people</u> a lesson that such a little brave girl should come walking from America & be willing to do more than they"

Of course Dad I know you will say that I can do with the money myself, yes, that is true, but also I get my living to a certain extent from the whole world so I must not grumble to give a part of it back again.

Now I must say good bye. Write me to this Hotel & if I have gone further it will be sent after me.

Your loving daughter

Lizzie xxxxx

Lizzie's father had obviously had enough of her constant demands for money. By this time he was living in Sowerby Bridge, near Halifax as his engineering business in nearby Luddenden had failed, although the exact circumstances are not known. In any event he probably did not have the same sort of financial resources as he had previously and this, coupled with his natural anxiety for his daughter, led him to plead with her to end her journey. His plea fell on deaf ears and Lizzie continued on westward stopping at Dresden and Leipzig before turning northwards to Wittenberg and then the German capital, Berlin. In each of these towns and cities she stopped a few days to lecture. From the newspaper reports we find that she now had a new agent, Sam Pearson, an American who, according to the newspapers, was an employee of the New York Polo magazine. He was able to speak German and translate for her, although his card, a copy of which survives in the globetrotter archives, simply says, "Sam Pearson, reporter and manager of Mrs Harry Humphries, 'The Globe Trotter', New York, USA". Later events, and the only surviving footage of Lizzie in which Sam Pearson appears, seem to suggest there was perhaps a closer relationship than merely agent and client, although it has to be said that it is pure speculation on my part. In Berlin she wrote another letter to her father.

Kaiser Hotel
Berlin W
1913 August 26th

My Dear old Dad

Why don't you write to me? What have I done to offend you? I have been expecting a letter from you for so long. I hope there is nothing wrong at home & that you are not in bad business worries. Have you heard from Percy? He came over to see me for 2 days only. I told them in the Hotel that he was my <u>brother</u>. I thought it was best because it would look rather queer for a gentleman to come from England to see me if he were not a relative, of course, if he could have spoken German then it would have been different, he is very much altered, & as you say I think it is for the better, he says that he is quite charmed with my voice now it is entirely different than in the olden days, more <u>Continental</u> Percy says, well Dad I guess it is.

I have been nearly half way around the world since the old Luddenden days & my <u>Yorkshire</u> dialect has vanished a little. Well Percy has promised to help me all in his power, he says that he thinks I am the <u>bravest</u> girl he has ever met & he can see that I am trying to do my trip <u>straight</u>. He says that he is getting a good salary nowadays & he can afford to help me to succeed &

that he will wait for me to marry him & I have <u>promised to be his wife</u>.

Now Dad maybe you are a little better satisfied but I should like to hear from you as soon as possible, write to me at once here & I will get your letter before I leave. I wonder if I will see you in Paris when I get down there & little Kelsey. I guess he is taller than I am now. What is he doing? Oh, I <u>wish</u> that he could be educated better. Percy would take him in hand if it could be arranged. Kelsey has brains but they need developing & now I see more & more every day of my life how valuable education & refinement is to a boy. Miss Conway also could do so much for him.

Good bye Dad & God bless you.

Your loving daughter Lizzie xxxx

If Lizzie could not understand why her father would be offended, she was being extremely obtuse. There was however better news. The mysterious Percy had come to see Lizzie in person and had taken the opportunity of seeing her face to face to ask her to marry him. She had accepted, presumably after he had offered her some financial help in her quest. Were these genuine offers or merely a ploy to take advantage of Lizzie? Lizzie was sufficiently appeased, however, to have their photograph taken, along with Vicksie the dog, at a photographic studio in the city and to send it home to her family at the same time as she sent the letter, along with the cryptic message,

"Do you know these two people? How could their pictures be taken together when one was in England and the other in Germany. 'Black Magic.'"

The Lady Globetrotter

Lizzie, Percy and Vicksie in Berlin

Whatever the reasons Lizzie was now on a high and she planned an audacious publicity stunt which could very well have ended in disaster. A grand autumn military parade was to be held in Berlin at the beginning of September and Lizzie rented a room in a hotel which was on the route. She knew that the German Kaiser, Wilhelm II, with his love of the pomp and ceremony of all martial matters, would be certain to participate. She ensured that she made herself and her quest conspicuous and as Wilhelm rode past she threw a bouquet of flowers at him from an upstairs window with a note attached "to the Great German Kaiser". She was immediately arrested but after her credentials were established she was released. In view of the tension which there was in Europe at the time between Germany on the one hand and France and Great Britain on the other, she was fortunate that she was not taken to be an assassin and dealt with more forcibly. However she had achieved her objective and the publicity reverberated as far away as America where the following newspaper report appeared in the New York Times.

SHE PELTED THE KAISER
Only with Flowers, but the Berlin Police Arrested Her

BERLIN, Sept 3. The police had temporary custody today of Mrs Harry Humphries, said to be an American, who yesterday threw a huge bouquet at the Kaiser as he was passing down the Friedrichstrasse to the military review at the Templehof field. She threw the flowers at him from her hotel window.

The woman had been making herself conspicuous around the hotel by wearing a placard bearing the legend, "40,000 days." When asked the meaning of it she said she intended wandering around the world in that number of days.

The bouquet thrown at the Kaiser bore a streamer inscribed "To the Great German Emperor."

The woman was released after her identity had been established and a note of it placed in the police records. She is considered highly eccentric but otherwise harmless.

Similar reports appeared in the German newspapers accompanied by stern warnings that security surrounding the Kaiser should be strengthened.

The Lady Globetrotter

*One of the publicity postcards which Lizzie sold
to help finance her journey*

Chapter 7
A Financial Crisis

Berlin to Dusseldorf
Copyright Google Maps ™ Mapping Service

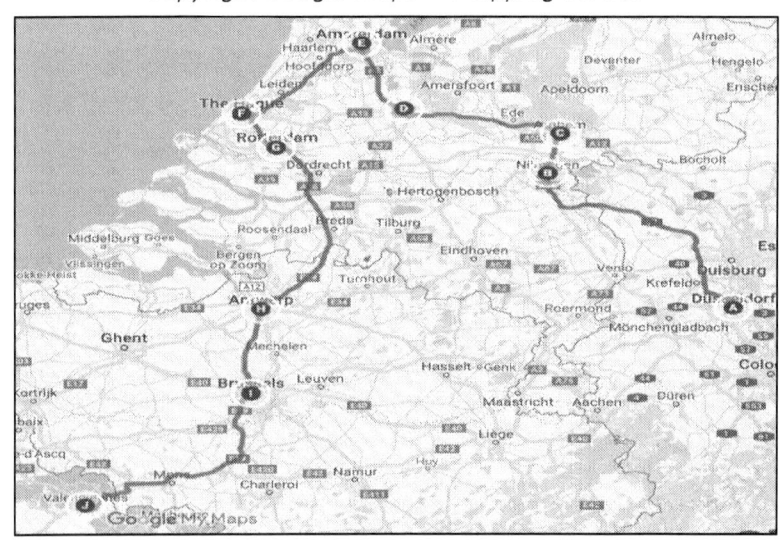

Holland and Belgium
Copyright Google Maps ™ Mapping Service

The Lady Globetrotter

No doubt buoyed up by Percy's visit and his proposal of marriage and by the success of her spectacular publicity stunt Lizzie continued her westward journey through Germany. Then she went into Holland accompanied, of course, by the usual fanfare of publicity, newspaper interviews, lectures and film shows. Unfortunately no letters from Lizzie have survived during this period from September until December. She then wrote home sometime in December, the letter is undated but, from its contents, we are able to determine that it was written towards the end of the month.

My Dear Dad

I don't understand why I never heard anything more about Theodore & Percy coming to see me at Xmas. I received your letter too late to reply to them but I had previously written to Theodore telling him to come or write to "Grand Hotel du Laboureur", Antwerp but I never received a line from him after your letter & I was so disappointed. I was showing my pictures in Majestic Theatre 20th, 21st, 22nd & 23rd but I asked the Director if I could show them on the 24th & 25th so that my brother could see them and he allowed me to do so but he did not pay me anything for those two days. So I worked 5 times for nothing & really I was so angry when they did not come because I knew I had a bad time in front of me & also I was very sick & should have stayed in bed those two days if I had not been expecting them.

It is very hard for me you know Dad. I never have a day when I can stay in bed & keep warm so that I can get better. A few weeks ago when I was in Holland I had to walk from The Hague to Rotterdam in rain. Snow & wind, it was terrible. I could hardly stand on my feet, I was out in the storm 7 hours & got wet to the skin. My boots which cost me nearly 60 marks in Berlin & which are <u>supposed</u> to be waterproof, were full of water & although I had on a new mackintosh & my skirt is made from pure leather, everything was wet through & I had to go straight to the Theatre to show my pictures & there I had to appear 3 times & stay till nearly 12 at night. Of course, the consequence was I took a chill & have been ill ever since but I could not stay in bed. One night I was wrapped in a net sheet from head to foot & then wrapped in blankets & had a hot drink to try & sweat it out but what was the use. I had to get up at 5 in the morning & then sail on a boat all day to a little island where I had an engagement to show my pictures twice that night.

I <u>have</u> to do all this in order to pay expenses & keep up my trip. Mr Pearson is doing better for me than anybody else I have had, except old Mr Hemfrey in Scotland (you remember the poor old man). If I can get my negatives alright from Stockholm & have some new copies made I can ask nearly twice the salary I am now getting from the Theatres. My Films have been going from bad to worse until the Director said they were not worth anything anymore.

If Percy had only kept his promise & loaned me the money to get my own machine I could have taken some beautiful pictures in Holland but now they are lost & I am afraid I shall never see the machine at all. My heart is very bitter, he said to me "I have <u>plenty</u> of money nowadays & I will help you to succeed". Never mind Dad you & I are not so selfish, thank the Lord.

Lizzie xxxxx

PS Will you please write to Mr Victor Meyer "Regina" Theatre Stockholm, Sweden & ask him, what is the matter there. I have written many, many times & received no reply & I am anxious. He has all my valuable papers & the negatives of my Cinematograph Films & I <u>must</u> have <u>new copies</u> made from them. The ones I have now are nearly worn out & I cannot show them much longer, it is very important. I have written again to him today. Please drop him a line at once. <u>Register</u> the letter so that he will be <u>sure</u> to get it.

So relations between Lizzie and her family, and Percy as well, seemed to have cooled somewhat. Her brother, Theodore, and Percy had not turned up for an expected visit at Christmas, somewhat to her annoyance. She was now crossing the Baltic to Sweden, a country she had previously visited, but the reason is apparent from the postscript to her letter. She was being let down by yet another man.

Lizzie's father, Alfred, and her brother Kelsey.

S/S "Rhea"
Malmo, Sweden
1913 December 31st

My Dear old Dad

I am writing to wish you & Kelsey a Happier New Year than the last one. I hope that God will not let you suffer as much in 1914 as He has this year. For myself I don't know what to say. I am afraid that there can be nothing for me for a long time only battle after battle to fight & I am beginning to think that there is no love in the world which is so true or self sacrificing as a Father's & Mother's.

I am disappointed in P ever since I was in Warsaw last June. He has promised, promised, promised, anybody can do that but "a friend in need is a friend indeed" & I am sorry to say he has not proved one. He has told you that I have not answered his letters but I have & I don't suppose he will show you the last one I wrote from Germany. I had to write to him just as I did because I was very angry with him for telling me such lies, but never mind in time everything will work itself out only he might have done so much for me & now I would be in a very different position today than I am.

My dear old Dad you will be very much astonished I know when I tell you that I am on board a Steamer going to Stockholm, Sweden again! I have come from Antwerp, Belgium. I left there on the 26th of Dec. & shall be in Stockholm on the 5th January.

My dear Dad I <u>had</u> to come, you know that I told you I left all my valuable papers & the negatives from my Films with Mr Meyer, I have written & written & written to him to send me new copies because my old one is spoilt, & I have had no letter from him since I was in Moscow. I am afraid that something is wrong, he treated me so kindly when I was in Stockholm, you remember I told you about it, so Dad I <u>must</u> come & see what is the matter & try to get my things back. I hope nothing has *happened to him & now I am almost without money in Antwerp. I had 300 francs & now I have only 52. I had to pay my Hotel bill in Antwerp & send Mr Pearson to Dresden in Germany until I come back & he had to take "Vicksie" with him because dogs are not allowed to come from Belgium to Sweden on account of a disease amongst dogs in Belgium & so Dad I feel so lost & lonesome on New Year's Eve.*

Unfortunately the remaining pages of this letter are missing so we don't know what else Lizzie said but it is clear that she felt let down by the mysterious Percy, and quite obviously no money had been forthcoming from him and once again she was short of finance. At the top of the first page she added the following.

S/S "Rhea"
Malmo, Sweden
1914 January 1st

Dear Dad

I did not post your letter last night, so today, the first day in the New Year, I can add you another line – now we are just leaving Malmo, the sea is thick with ice. We shall be 12 hours before we reach another port. It is a funny New Year's Day for me. I am in Sweden, my doggie in Germany, my Father & brother in England, my husband in America & I have no money in my pocket.

There then followed three very short letters, all from Sweden.

Stockholm
Sweden
1914 January 4th

My Dear old Dad

Here I am again & oh how tired & ill I feel after nine days on the water. I could have gone to New York sooner than I have come here but I am hoping that my long journey will not be all in vain. Oh how I am trembling for the result of my journey & my poor little doggie. I wonder if she is well. I hope she won't die for want of me, I have heard nothing of her for 9 long days & I am afraid she may be sick, nobody knows how much I think about her, she is everything to me & I cannot bear to think that something may happen to her while I am away.

I have just been telephoning to try & find Mr Meyer but it is Sunday a bad day to do anything & I can't go to the Post Office to see if there are any letters for me that is closed too. I shall have to wait till morning & now I can't write any more I feel too much upset about everything & the room is still going up & down just as if I was on the ship & soon I have to experience it all again & I know how bad it is, but tomorrow if I know something more I will write again so good bye for a little while & don't forget to tell me in your next letter why Theo & Percy did not come at Xmas. I don't think there will be time for me to get a reply here, but I will see tomorrow.

Stockholm
Sweden
1914 January 6th Tuesday

Dear Dad

I have seen Mr Meyer & my things are alright & he says he has written many letters but I have never got them, but I don't believe that he has written at all for a long time.

Anyhow I am so glad to have found them safe & now I shall put everything in the bank so that I know for sure it is quite safe & cannot get burnt up or anything happen to it & also I am glad that Mr Meyer has asked me to stay in Stockholm for a few days & show my pictures in the Cinematograph now I shall have some.

Stockholm
Sweden
1914 January 10th

A thousand thanks for the money you sent me. I <u>knew</u> you would not fail me if you could help it. Oh, I am so glad that now maybe I can send it back in a few days because I am earning some money here but I will wait until I get to Bruxelles & see if I can really do without it.

Now good bye. You can write me here to this Hotel but don't send more than <u>one</u> letter because I shall be going away soon.

Love to you & Kelsey

from your deeply grateful daughter,

Lizzie xxxxx

A sea journey at the height of winter obviously did not agree with her but she had managed to contact the elusive **Mr Meyer** and retrieve her belongings, in particular the films, and also managed to raise some much needed cash in the theatres and cinemas in Stockholm. Of course her long suffering father had also sent money to help her along. She then took a boat from Stockholm back to Antwerp arriving on 23rd January.

By the 27th of the month Lizzie had reached Brussels where she wrote an emotional letter to her father.

Grand Hotel Cosmopolite
Bruxelles
1914 January 27th

My Dear old Dad

Many thanks for your letter which I received in Antwerpen. I hardly know how to thank you for everything but I do very heartily. I have had such a bad time here. I got to Stockholm yes, but I had to wait 5 days before I could begin to earn one penny & then nearly all the money I got went for expenses & instead of paying me 50 Kroner for 3 days they paid me 25 & also spoilt my Film. Now when I have come back here to Bruxelles the theatres won't have my pictures because they have big holes in them. They are terrible & I think Mr Meyer has done it on purpose for revenge because I took all my papers & negatives away from him & now it is another struggle for me.

Last night Mr Pearson could have made an engagement for me here in Bruxelles for 500 francs for seven days but when the Director saw my Films he would not have them they are so bad, ruined almost. I cannot buy new copies now. I have all my negatives & with money of course can buy new copies but where on earth am I to get the money, if P had only done what he promised me so many times (to loan me £50 in a lump sum) then I would be earning very good money, but my hopes are gone as regards him.

Dad it is <u>absolutely necessary</u> that I have about £50 somehow. I <u>must</u> have it or else I shall have to give it up after all this hard work but I could <u>never</u> come back to England & oh, how much I want to do so some day & to help you as you have helped me. I tell you Dad if I can only <u>finish</u> my trip I shall be very rich after, there is no doubt about it & as I have said so many times before, you shall reap the benefit also. I have been wondering oh so long how I could get £50 & I think I see a way now if you will only help me & I pray to God that you will.

In your last letter you write to me asking that I shall sign a paper to show that you & not Theo shall draw my interest from Grandfather's property & that gave me an idea. <u>Could</u> you or <u>would</u> you ask Mr Jackson <u>to loan me £50</u> on the strength of what I shall receive some day when everything is straightened out & he can take what interest he likes. He <u>cannot lose</u> anything by it & he will have my life long thanks. It seems cruel of me to ask you this just when I could sign for <u>you</u> to take my interest Dad, but believe me it would set me on my feet so good & I could then send you at intervals all the money you have loaned me & you would then get it just the same as if I signed for you. Dad <u>don't</u> think I am talking wildly, you <u>know</u> that <u>sometime</u> I shall be entitled to more than £50 from the property & Mr Jackson knows it too & maybe then when I am old I won't need it & now I need it so much. If you were here & I showed you my Films & you also heard what the managers say, you would be convinced.

My dear old Dad don't you think that my Mother would be pleased to know that the money which <u>should</u> have been <u>hers</u>, helped me to be a success in what I am attempting to do. Yes, I am <u>sure</u> of it & I am also <u>sure</u> that you will go to see Mr Jackson for me at once & beg of him for the sake of my Mother, whom he respected, to help me, it is the last thing I can think of. In two weeks I ought to be in Paris & I <u>must</u> have new Films made before then or else no theatre will have me. Oh, if I could only <u>convince</u> you of the urgent need of this, but I think you know it already only it never occurred to you to ask Jackson & it did not to me before either.

I am enclosing a little note for Mr Jackson & when you have read it please give it to him & I hope he will answer straight away. I am sure he is a gentleman. I only saw him once but I remember he gave me that impression. If he is willing to help me Dad, give him the slip of paper where I sign for the money and now good bye.

Mr Pearson & Vicksie are quite well but Pearson is very much worried on my account. Poor boy he has gone to another town to try & arrange some business for me but I am quite sure that if they <u>insist</u> on seeing my Films before the contract is made, they won't have them at all & then I am left without money almost again.

Give my love to Kelsey & also a lot keep for yourself.

Your loving daughter

Lizzie xxxxx

Address Poste Restante Mons, Belgium

It was quite clear that Lizzie was now in dire financial straits and she laid the blame on Percy for not loaning the money she said he had promised, and Mr Meyer who she claimed had deliberately damaged her films, although the alleged grounds for doing this seem to be rather dubious. It is also clear that she was not above emotional blackmail when she invoked the memory of her mother. It is not known who Mr Jackson was and what relationship he had with the family. Her grandfather had died in 1883 leaving less than £70 in his will, which in today's money is less than £6,000 so one would have thought that any interest on Lizzie's share of that amount would be negligible.

Lizzie's plight was so serious that Sam Pearson was dispatched to England to speak to her family and to try to raise money. On his return he wrote back to Lizzie's father.

Grand Hotel Et Du Commerce
Valenciennes[1]
914 February 14th

My dear Mr Yates

I came back alright but – without any money. My friend was already gone. Another friend gave me just enough for the ticket.

I thank you ever so much for all your kindness and all you did for me. Crossley is the greatest blackguard I ever saw in my life and it's sure, he will be once very sorry for it!

Again, my dear Mr Yates, may God save & bless you, may better days come and may we see us again under better circumstances and in the best of health.

I am with heartiest greetings

Yours ever

Sam Pearson

On the same day Lizzie also wrote home.

Grand Hotel Et Du Commerce
Valenciennes
1914 February 14th

My Dear old Dad

I have received the money alright 628 francs & 75 centimes & I can never thank you enough for it. I shall hope to be able to show you my gratitude in a more substantial way than on paper before long. You must excuse my handwriting Dad, I can't help it my hand is trembling so much. I think it is the reaction after these last 14 days of suspense. Maybe I shall feel better in a few days when I have started working again.

Mr Pearson is back again & I am very glad though he looks very ill & worried. I have been very uneasy about him because I know he went to Crossley intending to _____ him, you understand, but now Pearson is back again & I sincerely hope that everything will go alright. The telegram which came with the money says "balance in fourteen days". I hope you will be able to send it then Dad because I shall then be in Paris & that is the place where I can get everything I need in Cinematograph business.

Now thank you for everything, the soap & "Mackintoshes Toffee" & give my love to Kelsey. I will write again soon. Crossley has sent <u>nothing</u>, he is a perfect Liar.

Yours with love,

Lizzie xxxxx

Pearson's quest had been unsuccessful in one respect but yet again Lizzie's father came to the rescue. The Fr.628 was approximately half of what she had asked but we do not know if he had heeded Lizzie's plea to obtain the money from her grandfather's estate or whether he had loaned the money out of his own personal savings.

One fact did emerge from these two letters, the surname of the enigmatic Percy. Quite what Sam Pearson intended to do to him is not clear as Lizzie simply put a line in her letter when describing his intentions. It seems however that they did meet and one can only wonder how that went.

We now know that Lizzie's fiancé from her Luddenden days was a gentleman called Percy Crossley, and there was a man of that name who lived in the village at the same time as she did. He was born in 1881 in Midgley, the adjoining village. He became the assistant teacher at Midgley school before moving onto Old Town School where he was the headmaster. By the time of the 1911 census the family had moved to Spring Hall Lane, Halifax. Sometime shortly after this the entire family, Percy's father, Thomas, his mother, Mary and his brother Jesse all moved to Mayland Terrace, Ben Rhydding, Yorkshire where Percy took up the post of headmaster at Ben Rhydding Council School. He married fellow teacher, Lucy Mann, in 1919 and continued living in the area until his death in 1959. As far as we can ascertain the couple had no children. Unfortunately we do not have his side of the story as I am sure it would make most interesting reading, but it is clear from the contents and tone of the letters of both Lizzie and Sam Pearson that they had a very low opinion of him. As far

The Lady Globetrotter

as anyone is aware there was no further contact between Lizzie and Percy.

Another of the publicity postcards sold by Lizzie

Chapter 8
Paris and the End

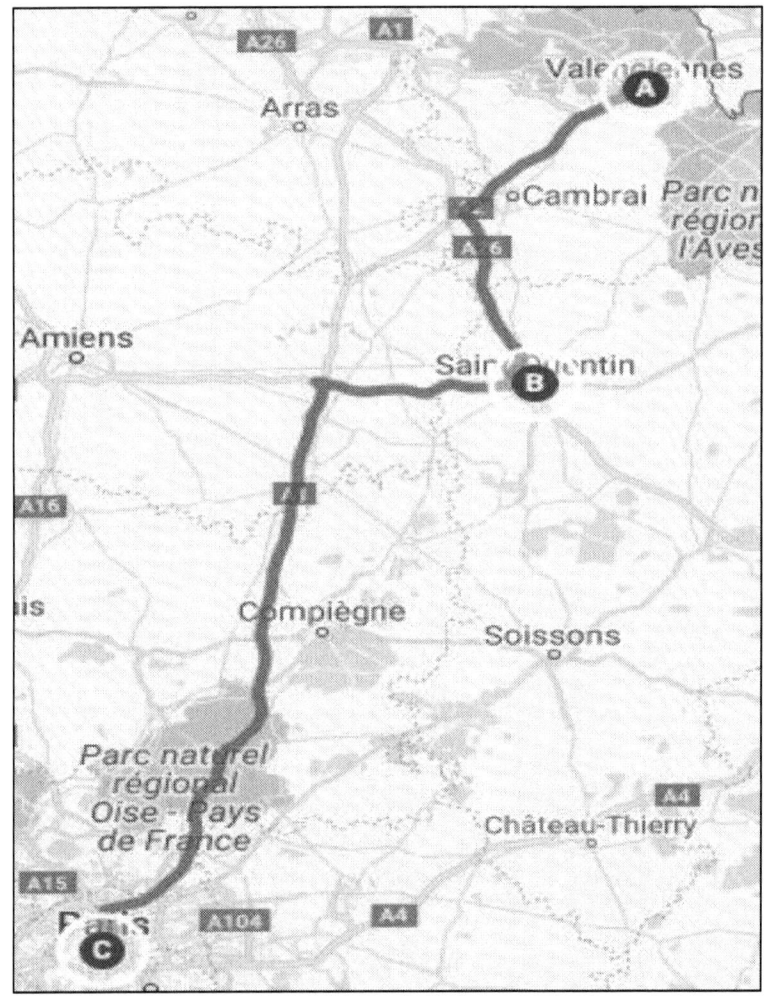

Valenciennes, France to Gay Paris
Copyright Google Maps™ Mapping Service

From newspaper articles we know that Lizzie was in St Quentin, France on 21st February 1914 and on April 6th she wrote home from Paris although the content infers that she had already been in the city some time.

Daily Mail Information Bureau
12 Boulevard Des Capucines
Paris
1914 April 16th

My Dear old Dad

Today I feel very sad it would have been Oswald's birthday you know if he had lived poor boy & I have been thinking about him all day. But he is a thousand times better off than we are Dad who have to earn our daily bread & earn it pretty hard too. I hope you could afford to put a few flowers on the grave for Easter. Tell me where have you been for this holiday with Kelsey?

It has not been at all like Easter for me. I have been very, very busy getting all my papers together & sorting & numbering them & making notes for my book. You must not be cross with me Dad because I am so long here in Paris, I cannot help it. I have been waiting for a chance to show my Cinematograph pictures in a theatre here & I am glad to say that in a few days I shall do so. You see Dad I must show them in Paris because it will make such a difference in my future life when I have finished my trip to be able to say & show a certificate that I have been engaged here. I can earn a lot more money & it is for your sake & for Kelsey as well that I am trying so hard.

You have been so good to me that I want to repay you all I can in the long run. You have no idea how much I love you both, more than ever have I realized it since I have been away from home. I only wish I could be with you for a few days & have a little time together. I think Kelsey would be glad too.

PC (Percy Crossley) promised that he would look after Kelsey's education & I was so glad but it is all a <u>lie</u> he wouldn't lift his finger for him but <u>I</u> will as soon as I have the chance. Don't be afraid Dad I will look after Kelsey & maybe you & he <u>can</u> go away from dirty, smoky old Sowerby Bridge to a much freer healthier country than you are now in, somewhere where the sky is always blue & where the people are not so narrow minded as in England. I know it looks to you now Dad as if I shall never finish my trip in time but I know what I am doing. In lots of places I shall have to do nothing only walk, walk, walk from day to day & then I shall make up my time so don't worry.

Now I must say good bye, give my love to Kelsey & my regards to Mrs Smith. Mr Pearson is still with me & he wishes to be kindly remembered to you. You can write to me Post Restante Central Paris, until I tell you different.

Your loving daughter,

Lizzie xxxxx

Lizzie's father was clearly exasperated with her and just as clearly Lizzie was attempting to placate him. What is amazing is that Lizzie still believed she could complete her journey in time. She had around two-thirds of her journey to complete across some of the most inhospitable terrain and through some of the most extreme climates that could be imagined. As well as having to walk down the west side of Africa she then had to turn north, trek the full length of the continent before turning eastwards across Asia, then to Australasia before crossing the Pacific to South America and northwards to New York. It is impossible to imagine that she could hope to complete this arduous journey by July 1915 even if she was walking most of the time without having to stop to give her lectures and film shows. In any case how would she manage the journey without sufficient funds unless she expected her father to continue subsidising her.

Lizzie's financial woes were now becoming desperate and, in an attempt to put them behind her once and for all, Sam Pearson returned to the Polo Monthly Magazine offices in New York to ask for sufficient funds to allow her to complete her endeavour. There were no transatlantic air flights in those days and Pearson had to take a passenger ship across the Atlantic, a journey of several weeks at least, so he could not be expected to return for some time. In the meantime Lizzie remained in Paris, unable to earn any money and relying entirely on the generosity of her father as can be seen from the letters she wrote to him.

Alliers - Post Restante
Paris Central
1914 May 6th

Dear old Dad

I received your letter & PO for one pound alright. I went straight to American Express & signed the cheque for you. I hope they sent it back alright. Mr Pearson is going from Paris on <u>Sunday</u> to Cherbourg & will sail for New York on Monday.

My dear old Dad I thank you ever so much for the promise to send me a pound a week until Pearson returns, which I hope will be soon. But Dad <u>can</u> I ask you to send me <u>this weekend so that I can have it by Saturday</u>, £2 so that Mr Pearson can buy for me a stock of things which I shall need, tea, cocoa, sugar, potatoes, rice etc. You see Dad I could never in this world live here in Paris for £5 a month if I had my meals in the Hotel. I have to cook them every one myself so I can live cheap. So I <u>beg</u> of you Dad to send me the £2 by Saturday, it will be much better for me because if <u>I</u> have to buy the things I shall be so much cheated, & I can't afford <u>that</u>. I have to pay 15 francs a week for my room so that will leave me only 10 francs for my meals for all the week. I hope Dad you will send me the money.

The Lady Globetrotter

Now about Aunt SJ you seem to think I asked her for money, <u>no,</u> I didn't. I only said to her that I couldn't afford to go over & see her unless she sent me the money for my fare. About the letter from Meyer in Stockholm, don't worry about it, he says I earned a lot of money there. Yes, I did but what about all my expenses there & back & all the time I lost money also, he does not say anything about that I guess. Of course, I may drop him a line sometime but that is not at all important & now Dad good bye & many thanks. I shall expect to hear from you on Saturday if possible.

Love to you & Kelsey

From Lizzie xxxxx

PS Yes it is better when you make the PO (Postal Order) *payable to Mrs E Humphries.*

As can be seen Lizzie's father was suspicious about her activities and how she was spending his money, Lizzie tried to explain it all away while at the same time asking for additional funds. Probably as a result, Sam Pearson took the time to write a short note to Lizzie's father, no doubt trying to allay his fears. Aunt SJ was probably Sarah Jane Thomas the sister of Lizzie's mother.

1914 May 6th

Dear Mr Yates.

It is a great relief to know that you are so prompt and good to help Lizzie. I shall try to get back as soon as possible, so that she can go on further and not be any trouble and worry to you. I hope everything with you is all right.

With heartiest Greetings to you and Kelsey

I am yours very truly

Sam Pearson

On the same day Lizzie also wrote home.

American Express Co
Visitors Writing Room (Not Official)
11 Rue ScribeParis
1914 May 17th

My Dear old Dad

I received the money yesterday & thank you very much. I am so sorry you have that old enemy again Influenza. I hope you will soon be better again, don't give in Dad because brighter days will come soon, I am sure, then you shall not have to worry about how to make ends meet. Maybe you have found my little articles of jewellery now they <u>must</u> be there, I gave them to you the night I left Scotland to go to Norway. They are amongst the papers, there is also a silver mesh purse a large one & some moon stones & a large brilliant etc. Also the present I got for my birthday in Aberdeen from the manager of the Cinematograph Theatre, it is a Scottish silver dirk set with two kinds of polished granite green & red. I remember it all so well Dad maybe you have never undone the package at all. There is also my old paisley silk shawl but don't worry about it Dad you are sure to find them.

Now I want to ask you what you would like to do in about a years time if everything goes alright & we are both spared. Would you like to live in England with Kelsey just like now & have a comfortable <u>business</u> of your own? or enough money to keep you & Kelsey in comfortable circumstances <u>without working</u> any longer? or would you prefer coming to live with me in New York of course I meant Kelsey also. I will make my home in America Dad it is much brighter and cleaner and more wholesome life than in England. I would dearly love to have you and Kelsey with me, don't think I am tricking with you. I mean all I say. I would like you to think it over and tell me what you think, of course Kelsey should have a much better

chance in America. Now I will say good bye, hoping in a few days you are better. I am enclosing a letter for Kelsey, will you send it to him.
Your loving daughter.
Lizzie xxxxxx

Lizzie's father was once again ill, he seemed to regularly have bouts of influenza. Lizzie, however, was bright and breezy and making plans for her future in America hoping to persuade her father to join her with her younger brother, Kelsey. She also took the time to write to him.

Poste Restante
Central Paris
France
1914 May 17th
My Dear Brother Kelsey
I thank you so much for writing to me, I got the postal order alright & I am very glad for it but sorry that Father is ill with Influenza. You do not tell me anything at all about yourself, what are you doing all day long while Father is away? what does Mrs Smith give you to eat? who are your friends now, do I know them? do you go to Chapel & Sunday School at all? what do you want most to be when you are grown up? I am asking you such a lot of questions I know Kelsey, that is because you never tell me anything about yourself & I would like to know.

Please write to me here again & tell me everything. I would be so very glad if I could see you again. Give my love to Martha Jane & Mrs Lumb. I am sending you this letter on my Russian note paper I thought it would be interesting for you but I don't think you can read it. I have put a mark under my name in Russian.

Write soon there's a good boy.

Lots of love from your Globe Trotting sister

Lizzie xxxxxxxx

PS

My Dear Kelsey I have walked through all these countries I have marked.

New York USA Canada New Foundland Prince Edward Isle New-Brunswick Nova Scotia England Ireland Scotland Wales Norway Denmark Sweden Finland Russia Poland Germany Austria Holland France

There was a gap of three weeks until Lizzie's next letter.

Paris
1914 June 6th

My Dear old Dad

I received the money last night Dad & I thank you ever so much. I am sorry now that I sent the post card but I was nearly crazy. I heard two men in the next room to mine speaking about Vicksie, they were saying she is a very valuable dog & they would try to steal her from me. I was so scared that I would have left the Hotel at once if I had the money, but when it did not come on Thursday I spoke to the wife of the proprietor & told her what I had overheard & that I was afraid they might try to open the door between their room & mine in the night & take Vicksie away, so she told me not to be afraid because she would make the men take another room upstairs.

So I am glad, you would be surprised Dad what a lot of people wanted my little dog in Russia. I nearly had to shoot one man who tried to steal her & in Germany an Officer offered 700 marks for her & here in France I have been asked to sell her many times.

The Lady Globetrotter

She is so much like a fox I have never seen another dog like her, even the cabmen here in Paris say "Renard" which means "Fox", when I pass by. I always keep her fast on a chain but Dad I <u>could</u> not sell her any more than you would sell <u>Kelsey</u>. She is all I have here to care for when you & Kelsey are so far away.

Now Dad I beg of you not to be so hard on me. I <u>know</u> that I have caused you more trouble than all the others put together but also Dad I have <u>tried</u> to do & <u>will</u> do <u>more</u> for you than all the others put together, as you know Mr Pearson has gone to "Polo Monthly" to try & get sufficient money for me to go ahead & try to finish this trip on time <u>but</u> if he <u>fails</u> to get money from them I think I shall <u>give it up</u>. People will laugh at me & say "I told you so, she couldn't do it" & that would be very hard to bear Dad because I <u>can</u> do it.

<u>If I have some money</u> to help me over the rough places, but never mind about that. If I <u>have</u> to I will put my pride in my pocket & come over to England & start to work in earnest by showing all my Films & speaking about my trip as far as I have gone. I shall get money enough to be able to send you some regularly Dad. Please don't worry & for Gods sake don't say I have driven you to the last ditch. The £50 you sent me here in France has <u>not</u> been squandered Dad, most of it has been used for my Film to put it in order. I ask you once more Dad to try & be hopeful for my sake & yours & Kelseys. Very soon now things will be different & for the better either one way or the other.

I have written to Pearson to cable me some money if he can't get back just yet, so that I don't have to go into your pocket. I will only ask you to help me next week with 20/- as usual & for you to send it so that I get it Friday morning. If Pearson cables money before then I will let you know.

Love from Lizzie

Lizzie had sent at least one postcard home and her father had obviously written to her giving her a piece of his mind which had the effect of an admission from Lizzie that she would give her journey up if Sam Pearson was unable to get any money from the Polo Monthly Magazine. She then wrote again.

Central Paris
1914 June 14th

Dear old Dad

I received your letter & PO for 20/- & I thank you very much. I have not received anything from Pearson yet but I am expecting a Cable every day now. I feel a little bit uneasy of course you can understand that he has taken with him most of my valuable signatures from everywhere I have been so that "Polo Monthly" can see for themselves that I have been doing my trip fair & square.

You know Dad as you say there are <u>such</u> a lot of shipwrecks it makes one anxious for those who <u>have</u> to cross the ocean & now I will tell you that I have been on the Swedish Steamer "<u>Storstad</u>" the trading vessel that ran into the "Empress of Ireland" & sent her down. It was in Sept 1911, HH & I sailed on the "Storstad" as guests of Captain Anderson from Sydney, Nova Scotia to St Johns, Newfoundland and we had awful rough weather, & when we came back from Newfoundland we sailed on a Norwegian ship called the "Kamfjord" & we were lost on the water for 2 days but afterwards got safe to land, but since then the "Kamfjord" has gone <u>down with all hands on board</u> & the Captain was <u>such</u> a nice man. He gave HH & I his own cabin because it was better than the others. It makes me feel very sad when I think of it & it is rather strange that Auntie Pollie is going back on the "Alsatian" she is the ship that brought some of the "dead passengers from the Empress of Ireland". I enclose a clipping from the "New York Herald", Paris edition, so you

The Lady Globetrotter

see both "Auntie Pollie & I are a bit connected with the great disaster. I hope that Auntie will get safely across, we have to trust in God that is all we can do. "What <u>has</u> to be <u>will</u> be". I wish Auntie Pollie would send me a line but maybe she is cross because I have been such a trouble to you.

I thank you very much for the "Courier ". Yes, I know that you would rather God had taken your life instead of Oswalds but <u>He</u> knows best, but what would have become of poor little Kelsey? I was in Russia & could not easily have got home & also I did not have enough <u>Film</u> to be able to show it in England & so make money to keep Kelsey, Oswald & I but now I <u>have</u> enough & if things go wrong with New York I shall come to England & make money. I only wish that Mother & Oswald were here to join with us, but God knows best why I also have had to suffer so much in my <u>private</u> life I don't know.

I will write again immediately Pearson replies.

Love to you & Kelsey & Auntie Pollie

Lizzie xxxxxx

There was still no word from Sam Pearson and it is apparent that Lizzie and her father were becoming concerned that perhaps some maritime disaster had befallen him. The 'Empress of Ireland' was a passenger ship which had sunk near the mouth of the St Lawrence River on the 29th May 1914 after colliding with the SS Storstad. She foundered in only 14 minutes and over 1,000 people were lost, an event which could only have fuelled their fears. It has not been possible to trace who Aunt Pollie was, neither the Yates nor the Kelsey families had relatives of that name. From what was said in the letter it seems that it was not only her father who was annoyed with Lizzie but also the rest of the family. A further letter followed on 6th July.

Central Paris
1914 July 6th

My Dear old Dad

My greatest thanks for the PO again. I have not been well these last few days. Yesterday I was in bed all day. I have an awful sharp pain in my side, that has been nothing new to me since I was in Russia but sometimes it is worse than others & this weekend it happened to be so, but don't be afraid Dad it will go away again. I don't half believe the doctors you know.

I had a letter from Auntie Pollie on Friday too. She says she is not cross with me but she would like to see me in a different position & so would I Dad for your sake. I quite agree with you that your own line of business is best & that you can make much more profit than selling a 1d article over a counter but Dad what will Kelsey do if he ever has to earn any kind of a living for himself? Has he got any <u>special</u> inclination for any line of business? Of course <u>I</u> shall look after him as well as I can if anything should happen to you which, God forbid, but Kelsey always seems like the "baby" of the family to me, I cannot think about him as grown up. What ambition has he got? I have asked him so many times to write & tell me what he would like to do but he <u>never writes</u> maybe you can tell me Dad. He has confided in you no doubt, & now I will say good bye, my next letter shall be longer.

I shall write to Auntie Pollie, she gave me her Canadian address & said she would like to hear from me.

It is too bad about T N Helliwell, these electrical storms have been something terrible here in Paris as well. I am glad that <u>you</u> have escaped them, be careful. Please give my love to Kelsey & keep some for yourself & tell Mrs Smith I hope she is well.

Your loving daughter

Lizzie xxxx

Although there is no mention of it in the letter momentous events were happening in Europe. On 28th June 1914 the Archduke Franz Ferdinand, heir to the Austro-Hungarian throne, had been assassinated in Sarajevo by Serbian nationalists. This act set in motion a series of events which culminated in one of the greatest tragedies to befall Europe. Lizzie, along with most of the rest of the continent was blissfully unaware of the tragedy which was about to unfold and she wrote cheerily home.

Hotel Des Pays-Bas
Paris
1914 July 11th
Dear old Dad

Thank you ever so much for the PO (Postal Order) *& letter again. I have had word from Mr Pearson he is coming back by way of Germany & will be here about the 25th July this month. I do not know what arrangement he has made with "Polo Monthly". I am very curious to know, I hope it is good news he brings.*

Anyhow Dad I am sure you will not have to send me any more money <u>after next week</u>, only one more PO oh I am so glad. I have been <u>such</u> a drag on you now it will be my turn to do something for <u>you</u> & Kelsey. I am sorry he has not been well. Yes the motor business is <u>fine</u> if he can learn it good. Will it cost a lot of money for him to become an apprentice? If you went to Southport to live would you take all your furniture from SB & also Mrs Smith to be housekeeper for you?

I am feeling better now thank the Lord. I am busy clipping all my newspapers & making a journal, it is so much better than having the <u>whole</u> newspapers, the freight is so much on my baggage.

Now Dad can I ask you to please post my money on <u>Wednesday</u> next week so that I get it <u>sure</u> on <u>Thursday</u> morning? I shall be so glad if you will do this for me & it is the <u>last time</u>.

Your loving daughter

Lizzie xxxxx

The last sentence that Lizzie wrote was prophetically true, but not for the reasons that she thought. She was in Neusse, Germany when war broke out between the United Kingdom and Germany on 4th August 1914 and she was forced to flee to avoid internment. She beat a hasty retreat to Antwerp. There she had to leave her luggage and beloved dog Vicksie, her companion of so many miles and adventures, before taking passage on the SS Montrose to England and her hometown.

It was a dispirited figure that trudged wearily into Halifax. Mrs Harry Humphries – The Lady Globetrotter, had been forced to give up her quest to walk around the world and prove a woman's endurance, at least for the moment, and her mood was as gloomy as the war clouds which were now gathered over the continent. Who would have thought that the assassination of a mid-European aristocrat in the Balkans would lead to a major European conflict and that Great Britain would become involved? To add insult to injury when asking for directions to the tram stop for her family home in Sowerby Bridge from a policeman, she had been viewed with suspicion owing to the accent she had picked up on her travels, no doubt due to the spy hysteria that was sweeping the country. A few hours later she was at her father's house in Burnley Road, Sowerby Bridge, back where it had all started.

Chapter 9
What Happened Next

After Lizzie's return home her story becomes fragmented and there are long periods when her whereabouts and activities are unknown. During her travels she had always claimed to be an American citizen but the records show that she only applied for American citizenship in 1959. Her application shows one surprising item of information. She had married Harry Norman on 3rd October 1919 in Galveston, Texas. I have been unable to find any record online of that marriage and there is no other mention of him in the archive that Lizzie left behind. In the application she says that he is deceased but once again I can find no online record of his death, moreover Lizzie travelled to and from America and Europe in the 1920s using her maiden name of Yates. I can think of no reason why Lizzie should fabricate a marriage but there is also no evidence to support it happening. Once again the story contains puzzling contradictions.

In October 1914 an American newspaper reported her arrival in Scranton, Pennsylvania accompanied by no less than her husband, Harry Humphries, and there is a tantalising record of the arrival in New York in September of Harry Humphries and his wife Elsie (a name which Lizzie often used) on the SS Philadelphia from Liverpool. The newspaper reported that she was walking across America to San Francisco and there is a publicity postcard in the archive advertising such a walk, the purpose of which was to raise funds for the Red Cross. Had there been a reunion back in England and the pair were now attempting another long-distance walk? After further newspaper reports of their progress in New Castle, Pennsylvania and Cleveland Ohio there is no further mention of this venture but we know that the reunion must have failed, perhaps spectacularly, as on 2nd May 1917 Harry married 17-year-old Hattie Kidd in Illinois. In 1918

America entered the First World War and Harry, along with all other eligible men, was required to register for the draft and from that we know that he now described himself as a painter, tellingly he only gives his mother-in-law's mailing address. He was probably travelling around the country and in September 1919 this headline appeared in the Arizona Republican newspaper.

TATTOOED MAN IS HELD WITH WIFE; CON GAME CHARGED

The 'tattooed man' was one Harry Humphries who, along with his wife Hattie, had been arrested for attempting a 'short change game' at a local grocery store. This entailed Harry purchasing something from the store, giving a five dollar bill in payment, leaving the shop and then returning a few minutes later claiming that the bill was of a larger denomination and he had been shortchanged. Unfortunately, and perhaps rather stupidly, he had already tried the trick in this store previously and the owner recognised him. This is undoubtedly our Harry as his wife described him as a one-time champion foot racer of the world and he himself admitted being in South Africa where he had secured a big diamond ring, stories he had always recounted to newspapers during his time with Lizzie. It had not been previously recorded that he was tattooed but apparently he was covered in them and the report describes them in some detail. The upshot was that Hattie claimed to have no knowledge of her husband's crime and walked free but Harry was found guilty and sentenced to between one and three years in Arizona State Prison. Hattie remained in Phoenix, Arizona and in November 1919 she made a plea to the state attorney general to release her husband in time for Christmas. She said that a miracle had been performed upon him while in prison. He had entered the prison in miserable health due to his morphine habit, which he had contracted as a result of illness, but had now gained some 40lb in weight. Presumably she believed he had now overcome his drug habit but this is not stated in the article. Her plea, however, fell upon deaf ears as in the 1920 USA census Harry was still residing in prison. Perhaps he was sewing mailbags, a fitting occupation for a master embroiderer. After this he and his wife disappear from the records and their fate is yet another frustrating mystery in this story.

However, I have almost certainly discovered the true identity of the elusive Harry. The draft card he completed in 1918 recorded his date of birth, 27th November 1881. In the first chapter I mentioned that Alfred Humphries' second wife, Anna Catherina, was herself a widow and had a son, Warnar Inch. After Anna's first husband died she returned to her home in Amsterdam, the Netherlands where she was duly recorded on the Amsterdam population register, along with Warnar and his sister Bertha. Warnar's date of birth was recorded as 27th November 1881. You would think that this is conclusive proof that he and Harry are one and the same person but, as usual in this most unusual of tales, there is a fly in the ointment. Warnar's birth certificate shows that he was born on 27th November 1880, the year before. In my experience however it is not unusual for people to misremember the year of birth of their children and it was not common practice for families to have a copy of their birth certificates in those days.

As for Lizzie, records show that she travelled back to Europe on several occasions, and she is known to have visited her brother Theo and his wife Ethel at their home in Sheffield in the 1920s. In 1923, along with American Scouts, she was in Holland at a gathering of the Scout movement. While she was there she walked from Amsterdam to The Hague and of course gave lectures in various places about her adventures as the Lady Globetrotter. In later life she worked as a commercial artist, but lost touch with her family and it was only after contacting a former neighbour that her brother, Theo, found that she had died a couple of years previously in 1960 in Dade County, Florida under the name Elsie A Norman.

The Lady Globetrotter

Lizzie with fellow scouts in Holland 1923

And so we end the incredible tale of The Lady Globetrotter and her attempt to walk around the world and prove a woman's endurance. Of course the main reason for Lizzie to embark on her quest was the $10,000 prize put up by the New York Polo magazine and the fame which would accompany such a feat. I do believe however that she was determined to cock a snook at early 20th century society with all the restrictions it placed on women. She was certainly a determined lady and like all people trying to place their mark on the world she could be self centered and selfish. In spite of all this I personally admire her grit and determination to complete her journey, unsuccessfully as it turned out, and throw off the shackles of such a restrictive society, and in this she certainly succeeded.

As for Harry what can be said. Blowhard, braggart, fantasist, compulsive liar and poser are some of the politer terms that could be used about him. The man had no moral compass and completely lacked any scruples. There are three recorded marriages for him, Amy Kiddell in 1908, Lizzie in 1910 and Hattie Kidd in 1917 and possibly another to May Humphries, the lady who was chasing him for money in San Francisco and with whom he had a son, and no doubt there were other relationships we do not know about, so he could obviously charm the ladies. Just as obviously he had a restless nature and tired of his wives and any venture he was involved in. The walk and the motorcycle journey spring to mind.

There are also several other records which have recently come to light and which probably involve him. In October 1897 Warner (surely a corruption of Warnar) Inch was convicted of common assault after shooting a young boy in the leg with a pistol. He somehow managed to convince the jury that he believed the firearm to be harmless and received a sentence of one day's imprisonment, the judge believing that *'the fear in the prisoner's mind as to the consequences of his rash action would be a sufficient warning to him.'* Then in 1899 one Harry Humphries born in 1881 in Bridport was listed as a deserter from the Wiltshire Regiment of the British Army. Perhaps this is where he got his tales of army life in the British Empire. Later, in 1909, Walter James Inch from Bridport smuggled three of his friends aboard the SMS North Point sailing from London to Philadelphia where they were discovered and listed as stowaways. Surely this is Warner using an anglicized form of his name. According to the ship's manifest he was returning to his wife in Pennsylvania

Although he was quite a reprehensible character and caused Lizzie a great deal of heartache and trouble it has to be said he has made the story that much more fascinating. It would be interesting to know what happened to him and how many more women he abandoned on his way through life.

Perhaps one day all the blanks in the lives of these two people will be filled in and their story completed. Who knows what other surprises may be found.

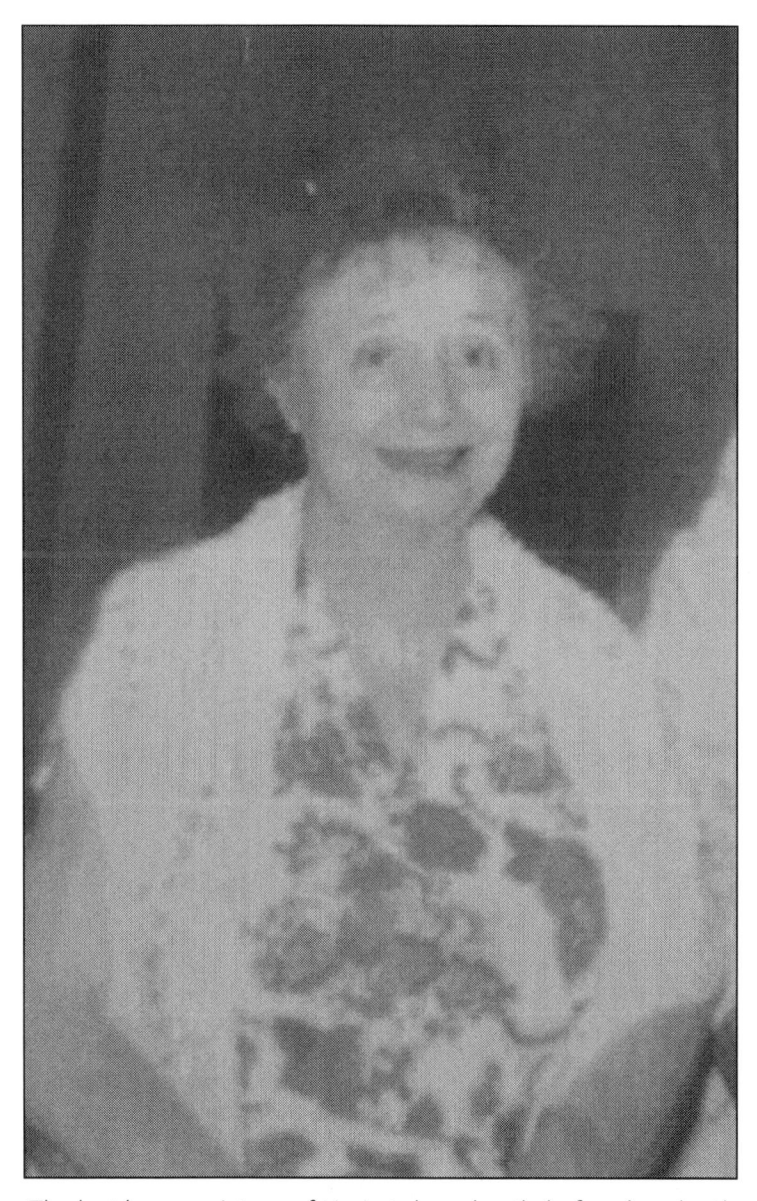

The last known picture of Lizzie taken shortly before her death

Printed in Great Britain
by Amazon